European Federation of Corrosion Publications
NUMBER 62

Testing tribocorrosion of passivating materials supporting research and industrial innovation: Handbook

Edited by
J.-P. Celis & P. Ponthiaux

EUROPEAN FEDERATION OF CORROSION
FÉDÉRATION EUROPÉENNE DE LA CORROSION
EUROPÄISCHE FÖDERATION KORROSION

Published for the European Federation of Corrosion
by CRC Press
on behalf of
The Institute of Materials, Minerals & Mining

 CRC Press
Taylor & Francis Group
Boca Raton London New York

CRC Press is an imprint of the
Taylor & Francis Group, an **informa** business

The Institute of Materials,
Minerals and Mining

First published 2012 by Maney Publishing

Published by CRC Press on behalf of the European Federation of Corrosion and
The Institute of Materials, Minerals & Mining

2 Park Square, Milton Park, Abingdon, Oxon OX14 4RN
711 Third Avenue, New York, NY 10017, USA

First issued in paperback 2017

CRC Press is an imprint of Taylor & Francis Group, an Informa business

ISBN 13: 978-1-138-11608-5 (pbk)
ISBN 13: 978-1-907975-20-2 (hbk)

ISSN 1354-5116

Contents

Series Introduction   *iv*

Volumes in the EFC series   *vi*

List of symbols   *xi*

0   Introduction
    J.-P. Celis and P. Ponthiaux   1

1   Phenomena of tribocorrosion in medical and industrial sectors
    J. Takadoum and A. Igartua   14

2   Depassivation and repassivation phenomena: synergism in
    tribocorrosion
    R. Oltra   29

3   Specific testing techniques in tribology: laboratory techniques for
    evaluating friction, wear, and lubrication
    S. Achanta and D. Drees   55

4   Specific testing techniques in tribology and corrosion:
    *Electrochemical techniques for studying tribocorrosion processes
    in situ*
    V. Vignal, F. Wenger and B. Normand   88

5   Design of a tribocorrosion experiment on passivating surfaces:
    Modelling the coupling of tribology and corrosion
    F. Wenger and M. Keddam   119

6   Towards a standard test for the determination of synergism in
    tribocorrosion: Design of a protocol for passivating materials
    N. Diomidis   150

7   Towards a standard test for the determination of synergism in
    tribocorrosion: Detailed testing procedure for passivating materials
    R. Bayon   167

8   Normative approach
    A. Igartua and M.P. Gomez-Tena   184

    Index   201

European Federation of Corrosion (EFC) publications: Series introduction

The EFC, founded in 1955, is a Federation of 33 societies with interests in corrosion and is based in twenty-six different countries throughout Europe and beyond. Its member societies represent the corrosion interests of more than 25,000 engineers, scientists and technicians. The Federation's aim is to advance the science of the corrosion and protection of materials by promoting cooperation in Europe and collaboration internationally. Asides from national and international corrosion societies, universities, research centres and companies can also become Affiliate Members of the EFC.

The administration of the Federation is in the hands of the Board of Administrators, chaired by the EFC President, and the scientific and technical affairs are the responsibility of the Science and Technology Advisory Committee, chaired by the STAC Chairman and assisted by the Scientific Secretary. The General Assembly approves any EFC policy prepared and presented by the BoA. The Federation is managed through its General Secretariat with three shared headquarters located in London, Paris and Frankfurt.

The EFC carries out its most important activities through its nineteen active working parties devoted to various aspects of corrosion and its prevention, covering a large range of topics including: Corrosion and Scale Inhibition, Corrosion by Hot Gases and Combustion Products, Nuclear Corrosion, Environment Sensitive Fracture, Surface Science and Mechanisms of Corrosion and Protection, Physico-chemical Methods of Corrosion Testing, Corrosion Education, Marine Corrosion, Microbial Corrosion, Corrosion of Steel in Concrete, Corrosion in Oil and Gas Production, Coatings, Corrosion in the Refinery Industry, Cathodic Protection, Automotive Corrosion, Tribo-Corrosion, Corrosion of Polymer Materials, Corrosion and Corrosion Protection of Drinking Water Systems, Corrosion of Archaeological and Historical Artefacts. The EFC is always open to formulating new working parties in response to the demands brought about by developing technologies and their ensuing corrosion requirements and applications.

The European Federation of Corrosion's flagship event is EUROCORR, the most important Corrosion Congresses in Europe, which is held annually in a different European country in September of each year. To date, 27 EUROCORR conferences have taken place in 12 different countries and they have gained a reputation for their high technical quality, global perspective and enjoyable social programme. Another channel for the EFC's valuable transfer of knowledge is the EFC "green" book series which are the fruit of the collaboration and high scientific calibre within and amongst the EFC working party members and are emblematic of the EFC editorial policy.

EFC Offices are located at:
European Federation of Corrosion, The Institute of Materials, Minerals and Mining,
1 Carlton House Terrace, London SW1Y 5DB, UK

Fédération Européenne de la Corrosion, Fédération Française pour les sciences de la Chimie, 28 rue Saint-Dominique, F-75007 Paris, France

Europäische Föderation Korrosion, DECHEMA e.V., Theodor-Heuss-Allee 25, D-60486 Frankfurt-am-Main, Germany

Volumes in the EFC series

1 **Corrosion in the nuclear industry**
   Prepared by Working Party 4 on Nuclear Corrosion*

2 **Practical corrosion principles**
   Prepared by Working Party 7 on Corrosion Education*

3 **General guidelines for corrosion testing of materials for marine applications**
   Prepared by Working Party 9 on Marine Corrosion*

4 **Guidelines on electrochemical corrosion measurements**
   Prepared by Working Party 8 on Physico-Chemical Methods of Corrosion Testing*

5 **Illustrated case histories of marine corrosion**
   Prepared by Working Party 9 on Marine Corrosion

6 **Corrosion education manual**
   Prepared by Working Party 7 on Corrosion Education

7 **Corrosion problems related to nuclear waste disposal**
   Prepared by Working Party 4 on Nuclear Corrosion

8 **Microbial corrosion**
   Prepared by Working Party 10 on Microbial Corrosion*

9 **Microbiological degradation of materials and methods of protection**
   Prepared by Working Party 10 on Microbial Corrosion

10 **Marine corrosion of stainless steels: chlorination and microbial effects**
   Prepared by Working Party 9 on Marine Corrosion

11 **Corrosion inhibitors**
   Prepared by the Working Party on Inhibitors*

12 **Modifications of passive films**
   Prepared by Working Party 6 on Surface Science*

13 **Predicting CO_2 corrosion in the oil and gas industry**
   Prepared by Working Party 13 on Corrosion in Oil and Gas Production*

14 **Guidelines for methods of testing and research in high temperature corrosion**
   Prepared by Working Party 3 on Corrosion by Hot Gases and Combustion Products

15   **Microbial corrosion: Proceedings of the 3rd International EFC Workshop**
Prepared by Working Party 10 on Microbial Corrosion

16   **Guidelines on materials requirements for carbon and low alloy steels for H₂S-containing environments in oil and gas production (3rd Edition)**
Prepared by Working Party 13 on Corrosion in Oil and Gas Production

17   **Corrosion resistant alloys for oil and gas production: guidance on general requirements and test methods for H₂S service (2nd Edition)**
Prepared by Working Party 13 on Corrosion in Oil and Gas Production

18   **Stainless steel in concrete: state of the art report**
Prepared by Working Party 11 on Corrosion of Steel in Concrete*

19   **Sea water corrosion of stainless steels: mechanisms and experiences**
Prepared by Working Party 9 on Marine Corrosion and Working Party 10 on Microbial Corrosion

20   **Organic and inorganic coatings for corrosion prevention: research and experiences**
Papers from EUROCORR '96

21   **Corrosion-deformation interactions**
CDI '96 in conjunction with EUROCORR '96

22   **Aspects of microbially induced corrosion**
Papers from EUROCORR '96 and EFC Working Party 10 on Microbial Corrosion

23   **CO₂ corrosion control in oil and gas production: design considerations**
Prepared by Working Party 13 on Corrosion in Oil and Gas Production*

24   **Electrochemical rehabilitation methods for reinforced concrete structures: a state of the art report**
Prepared by Working Party 11 on Corrosion of Steel in Concrete

25   **Corrosion of reinforcement in concrete: monitoring, prevention and rehabilitation**
Papers from EUROCORR '97

26   **Advances in corrosion control and materials in oil and gas production**
Papers from EUROCORR '97 and EUROCORR '98

27   **Cyclic oxidation of high temperature materials**
Proceedings of an EFC Workshop, Frankfurt/Main, 1999

28   **Electrochemical approach to selected corrosion and corrosion control**
Papers from the 50th ISE Meeting, Pavia, 1999

29   **Microbial corrosion: proceedings of the 4th International EFC Workshop**
Prepared by the Working Party on Microbial Corrosion

30   **Survey of literature on crevice corrosion (1979–1998): mechanisms, test methods and results, practical experience, protective measures and monitoring**
Prepared by F. P. Ijsseling and Working Party 9 on Marine Corrosion

31   **Corrosion of reinforcement in concrete: corrosion mechanisms and corrosion protection**
Papers from EUROCORR '99 and Working Party 11 on Corrosion of Steel in Concrete

32   **Guidelines for the compilation of corrosion cost data and for the calculation of the life cycle cost of corrosion: a working party report**
Prepared by Working Party 13 on Corrosion in Oil and Gas Production

33   **Marine corrosion of stainless steels: testing, selection, experience, protection and monitoring**
Edited by D. Féron on behalf of Working Party 9 on Marine Corrosion

34   **Lifetime modelling of high temperature corrosion processes**
Proceedings of an EFC Workshop 2001 Edited by M. Schütze, W. J. Quadakkers and J. R. Nicholls

35   **Corrosion inhibitors for steel in concrete**
Prepared by B. Elsener with support from a Task Group of Working Party 11 on Corrosion of Steel in Concrete

36   **Prediction of long term corrosion behaviour in nuclear waste systems**
Edited by D. Féron on behalf of Working Party 4 on Nuclear Corrosion

37   **Test methods for assessing the susceptibility of prestressing steels to hydrogen induced stress corrosion cracking**
By B. Isecke on behalf of Working Party 11 on Corrosion of Steel in Concrete

38   **Corrosion of reinforcement in concrete: mechanisms, monitoring, inhibitors and rehabilitation techniques**
Edited by M. Raupach, B. Elsener, R. Polder and J. Mietz on behalf of Working Party 11 on Corrosion of Steel in Concrete

39   **The use of corrosion inhibitors in oil and gas production**
Edited by J. W. Palmer, W. Hedges and J. L. Dawson on behalf of Working Party 13 on Corrosion in Oil and Gas Production

40   **Control of corrosion in cooling waters**
Edited by J. D. Harston and F. Ropital on behalf of Working Party 15 on Corrosion in the Refinery Industry

41   **Metal dusting, carburisation and nitridation**
Edited by H. Grabke and M. Schütze on behalf of Working Party 3 on Corrosion by Hot Gases and Combustion Products

42   **Corrosion in refineries**
Edited by J. D. Harston and F. Ropital on behalf of Working Party 15 on Corrosion in the Refinery Industry

43   **The electrochemistry and characteristics of embeddable reference electrodes for concrete**
Prepared by R. Myrdal on behalf of Working Party 11 on Corrosion of Steel in Concrete

44   **The use of electrochemical scanning tunnelling microscopy (EC-STM) in corrosion analysis: reference material and procedural guidelines**
Prepared by R. Lindström, V. Maurice, L. Klein and P. Marcus on behalf of Working Party 6 on Surface Science

45   **Local probe techniques for corrosion research**
Edited by R. Oltra on behalf of Working Party 8 on Physico-Chemical Methods of Corrosion Testing

46   **Amine unit corrosion survey**
Edited by J. D. Harston and F. Ropital on behalf of Working Party 15 on Corrosion in the Refinery Industry

47   **Novel approaches to the improvement of high temperature corrosion resistance**
Edited by M. Schütze and W. Quadakkers on behalf of Working Party 3 on Corrosion by Hot Gases and Combustion Products

48   **Corrosion of metallic heritage artefacts: investigation, conservation and prediction of long term behaviour**
Edited by P. Dillmann, G. Béranger, P. Piccardo and H. Matthiesen on behalf of Working Party 4 on Nuclear Corrosion

49   **Electrochemistry in light water reactors: reference electrodes, measurement, corrosion and tribocorrosion**
Edited by R.-W. Bosch, D. Féron and J.-P. Celis on behalf of Working Party 4 on Nuclear Corrosion

50   **Corrosion behaviour and protection of copper and aluminium alloys in seawater**
Edited by D. Féron on behalf of Working Party 9 on Marine Corrosion

51   **Corrosion issues in light water reactors: stress corrosion cracking**
Edited by D. Féron and J-M. Olive on behalf of Working Party 4 on Nuclear Corrosion

52   **Progress in Corrosion – The first 50 years of the EFC**
Edited by P. McIntyre and J. Vogelsang on behalf of the EFC Science and Technology Advisory Committee

53   **Standardisation of thermal cycling exposure testing**
Edited by M. Schütze and M. Malessa on behalf of Working Party 3 on Corrosion by Hot Gases and Combustion Products

54   **Innovative pre-treatment techniques to prevent corrosion of metallic surfaces**
Edited by L. Fedrizzi, H. Terryn and A. Simões on behalf of Working Party 14 on Coatings

55   **Corrosion-under-insulation (CUI) guidelines**
Prepared by S. Winnik on behalf of Working Party 13 on Corrosion in Oil and Gas Production and Working Party 15 on Corrosion in the Refinery Industry

56   **Corrosion monitoring in nuclear systems**
Edited by S. Ritter and A. Molander

57 **Protective systems for high temperature applications: from theory to industrial implementation**
Edited by M. Schütze

58 **Self-healing properties of new surface treatments**
Edited by L. Fedrizzi, W. Fürbeth and F. Montemor

59 **Sulphur-assisted corrosion in nuclear disposal systems**
Edited by D. Féron, B. Kursten and F. Druyts

60 **Methodology of crevice corrosion testing for stainless steels in natural and treated seawaters**
Edited by U. Kivisäkk, B. Espelid and D. Féron

61 **Inter-laboratory study on electrochemical methods for the characterisation of CoCrMo biomedical alloys in simulated body fluids**
Edited by A. Igual Munoz and S. Mischler

Symbols	Description	Units
A_{act}	Area of active material in the wear track	cm^2
A_o	Sample area	cm^2
A_{pass}	Area of material covered by a passive film	cm^2
A_{repass}	Area of repassivated material in the wear track	cm^2
A_{tr}	Wear track area	cm^2
B	Constant	mV
d	Density of bare material	$g\,cm^{-3}$
E_i	Elastic modulus of material i	GPa
E_{oc}	Open circuit potential	V vs. ref
F	Faraday constant	$96\,485\;C\,mol^{-1}$
F_n	Normal force	N
i_{act}	Corrosion current density of active material	$mA\,cm^{-2}$
i_{pass}	Corrosion current density of passive material	$mA\,cm^{-2}$
K_c	Ratio of the specific material loss due to corrosion over the specific material loss due to mechanical wear in the wear track	
K_m	Ratio of the specific material loss due to mechanical wear of the active material over the specific material loss due to mechanical wear of the repassivated material in the wear track	
M	Molecular weight	$g\,mol^{-1}$
N	Number of cycles	
n	Number of electrons	
P_{Hmax}	Maximum Hertzian contact pressure	MPa
R_{cb}	Counter body radius	mm
R_o	Polarization resistance of passive material	Ω
R_{tr}	Track radius	mm
R_1	Polarization resistance under continuous sliding conditions	Ω
R_{1act}	Polarization resistance of active area under continuous sliding conditions	Ω
R_{1pass}	Polarization resistance of passive and repassivated area under continuous sliding conditions	Ω
r_{act}	Specific polarization resistance of active material	$\Omega\,cm^2$
r_{pass}	Specific polarization resistance of passive material	$\Omega\,cm^2$
t_{rot}	Rotation period	s
t_{lat}	Latency time	s
t_{off}	Off time during intermittent tests	s
t_{reac}	Reactivity time to form surface film	s
W^c_{act}	Material loss due to corrosion of active area in the wear track	cm^3

W^c_{repass}	Material loss due to corrosion of repassivated material in the wear track	cm^3
W^m_{act}	Material loss due to mechanical wear of active material in the wear track	cm^3
W^m_{repass}	Material loss due to mechanical wear of repassivated material in the wear track	cm^3
W_t	Total material loss	cm^3
W_{tr}	Material loss in the wear track	cm^3
w^c_{act}	Specific material loss due to the corrosion of active material	$cm^3\ cm^{-2}\ cycle^{-1}$
w^c_{repass}	Specific material loss due to the corrosion of repassivated material	$cm^3\ cm^{-2}\ cycle^{-1}$
w^m_{act}	Specific material loss due to the mechanical wear of active material	$cm^3\ cm^{-2}\ cycle^{-1}$
w^m_{repass}	Specific material loss due to the mechanical wear of repassivated material	$cm^3\ cm^{-2}\ cycle^{-1}$
α	Contact radius	mm
μ	Coefficient of friction	
v_i	Poisson ratio of material i	GPa
$\sigma_{0.2}$	Yield strength	MPa

0
Introduction

Jean-Pierre Celis

Katholieke Universiteit Leuven, Dept. MTM, B-3001 Leuven, Belgium

Jean-Pierre.Celis@mtm.kuleuven.be

Pierre Ponthiaux

Ecole Centrale Paris, Dept. LGPM, F-92295 Châtenay-Malabry, France

pierre.ponthiaux@ecp.fr

This book is intended primarily for industrial laboratories and technical centres interested in standardised tests in the field of tribology and combined tribology–corrosion. In this context, besides the development of a test for measuring the susceptibility to tribocorrosion of passivating materials, the authors have sought, through this book:

-   to provide protagonists close to industry with a tool for dialogue and exchange with tribology experts and academic researchers
-   to clarify the approach underpinned by the development of a protocol on tribo-corrosion and its implementation 'ultimately' to give birth to an international standard.

The support of the European Federation of Corrosion (EFC) in the development of this book on tribocorrosion is most appreciated by the authors. Parts of this book were elaborated during different joint research and development projects such as COST 533 Biotribology funded by the EU, the EUREKA Umbrella ENIWEP, the EU Network of Excellence on 'Complex Metallic Alloys' (CMA) under contract no. NMP3-2005-CT-500145, and the Scientific Research Community on Surface Modification of Materials (WOG) funded by the Science Foundation Flanders.

Tribology aims at the study and scientific interpretation of experimental facts, but it also has a very specific purpose in all areas related to technology, namely the search for and coding of methods that allow to achieve a 'good behaviour' of moving mechanical contacts, knowing that parts constituting machinery 'fail' today more often by their surfaces than by their volume.

The knowledge on tribology has become a necessity for many reasons, namely:

-   to ensure the proper functioning and *reliability* of machinery
-   to reduce the cost of obtaining rubbing surfaces
-   to improve the *performance* and *longevity* of machines
-   to guarantee the safety of belongings and persons, particularly in transportation and in the health sector
-   to contribute to *public health* and comfort, e.g. by reducing the numerous *noise* sources linked to mechanical contacts.

Tribocorrosion is a subject of a rare universality that leads us to explore many aspects of science and technology, with of course friction and adhesion in the first instance, but also:

- *materials science* for obvious reasons
- *crystallography* because one must consider the properties of micro-crystals and their stacking to form surface layers on parts
- *physico-chemistry of surfaces* particularly with regard to the properties and composition of surface layers, the strength of their bonding to substrates, and their interaction with lubricants
- *thermodynamics* because the thermal phenomena, and in particular flash temperatures resulting from the interlocking between surface asperities, affect the behaviour of surfaces and their degradation by wear
- *thermochemistry* because these high peak temperatures cause severe chemical reactions at contacting interfaces
- *electrochemistry* since it dictates the behaviour of surfaces in contact with ion conductive electrolytes
- *mechanical strength of materials* especially in the case of contacts called 'occasional' or 'line networks', that can be drastically lowered not because of surface stresses, but owing to fatigue effects in sub-surface layers.

Areas which must be added are, without being exhaustive, intrinsic and extrinsic parameters linked to the system, namely contact stiffness, contact pressures, wear processes, lubrication phenomena, evolution of structural properties of materials, effect of heat treatments, as well as benefits and drawbacks of surface coatings and surface treatments in general.

0.1   From tribology to tribocorrosion

It is generally acknowledged that economic losses due to friction and wear account for 6% to 10% of the Gross National Product (GNP) of industrialised countries. Wear represents 30% of the causes of dysfunction of mechanical systems. The main causes of material degradation in engineering parts are summarised and illustrated in Table 0.1.

Wear and corrosion are two processes leading to surface damage resulting from a progressive removal of material as a result of, respectively, mechanical and electro-chemical processes. When these two types of degradation occur simultaneously, e.g. when two solid bodies are in relative motion while being immersed in an aggressive

Table 0.1 Main material degradation phenomena in engineering parts

Degradation phenomenon	Relative occurrence, %
Abrasion	30
Adhesion	15
Surface fatigue	15
Thermal fatigue	12
Contact corrosion (tribocorrosion)	10
Corrosion	10
Erosion, cavitation	8

environment, one calls this 'tribocorrosion'. To better understand the specificity of tribocorrosion, it is worth recalling briefly the difference between tribological studies and corrosion studies.

Tribology deals primarily with mechanical aspects related to friction and wear, and the effect of lubrication on these factors. It requires testing and data analysis with a multidisciplinary approach exploring many aspects of science and technology. In order to perform a tribological test adequately, information must be available on both the products and materials of interest to the industry, and on the type of contact as active in an industrial system. The knowledge required can thus, in the first instance, be listed as follows:

- related to the *product and materials of interest*:
  - mechanical properties of the surface layers present on each protagonist once steady operation is achieved. This includes data on hardness, H, elasticity limit, R_e, mechanical strength, R_m, elasticity modulus, E, shear modulus, G, residual stress, S_r, accumulated elastic energy, U_{el}, plastic deformation, σ, and fracture toughness, $K.\varepsilon$
  - surface properties such as surface energy, molecular adsorption, chemical reactivity, surface film formation (known as 'third body' whereas the contacting parts are known as 'first bodies'), and topography (roughness at different length scales).
- related to the *contact system*:
  - the nature of the contact being open or closed. This has a large effect on the escape or trapping of wear particles from the contact, and thus on the load distribution in the contact. In the case of lubricated systems, it affects the flow of liquid lubricants in and out of the contact zone
  - the characteristics of the tribological system considered as a whole. This includes the stiffness of the system inclusive measuring devices, the dimensional tolerances, elasticity, vibrations, etc., which result from the assembly of the various parts activated to achieve a relative motion.

Corrosion in turn is the result of physico-chemical (oxidation/reduction) processes taking place at the surface of materials (metallic, ceramic and organic). The study of these processes permits, through variations of the electrochemical potential and current, assessment of the reactivity of surfaces of materials with respect to their environment, and the evolution of that reactivity with time. Corrosion is always the result of a reaction due to heterogeneity, in its broadest sense, in a material or in its environment. It generates changes in surface composition which often result in a deterioration of the functional properties of materials (e.g. material loss, enlarged crack sensitivity), but also in a modification of the environment as such (e.g. contamination by reaction products, pH changes). Also here, the simulation of corrosion at the lab scale requires precise information on the materials involved in the industrial system and the environmental conditions, and their possible fluctuations during operation. Contaminants present in either the materials (e.g. trace elements) or in the environment (e.g. acid rain) can strongly affect the corrosion behaviour.

0.2   Definition of tribocorrosion and its importance

Tribocorrosion is defined as the study of the influence of chemical, electrochemical and/or biological environmental factors on the friction and wear behaviour of surfaces of materials in mechanical contact with each other, and undergoing a relative motion relative to each other.

The consequences of a coupling of friction with corrosion are nowadays still difficult to master. Indeed, the knowledge of the tribological behaviour of a material couple in the absence of any aggressive media, and the knowledge of the electrochemical behaviour in the absence of any mechanical impact, are not sufficient to derive the tribocorrosion behaviour of that system in which that same material couple would be used [1]. It has been noticed in many articulating systems that friction may modify the sensitivity of the constituting materials to corrosion, and on the other hand, corrosion may affect friction between moving contacting parts. Such a loading process may lead to a 'synergism' between electrochemical and mechanical processes [2]. This usually accelerates the tribo-chemical degradation of the material, and affects the coefficient of friction [3]. The complexity of a tribocorrosion system is given schematically in Figure 0.1.

The macroscopic quantities conventionally measured in a tribological test are:

- the coefficient of friction based on the friction force recorded *in situ*
- the wear rate determined *ex situ* from a loss of material on one or both contacting materials due to mechanical interactions leading to the formation and escape of debris
- the contact temperature recorded *in situ*
- the vibrations of the contacting parts, and
- the noise eventually emitted during the test.

In a tribocorrosion test, besides these macroscopic quantities, complementary *in situ* measurements of the electrochemical potential of the contacting surfaces immersed in an aggressive environment, and/or an *in situ* measurement of the corrosion current can be performed while the contacting parts are in relative motion. That motion can be a continuous or discontinuous one; it can be a unidirectional or a reciprocating one.

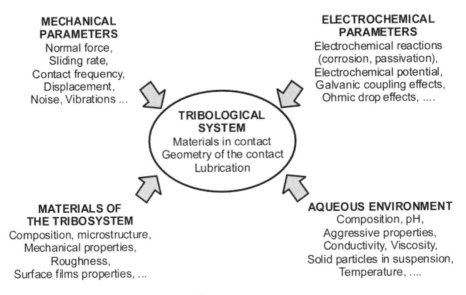

0.1 Complexity of a tribocorrosion system

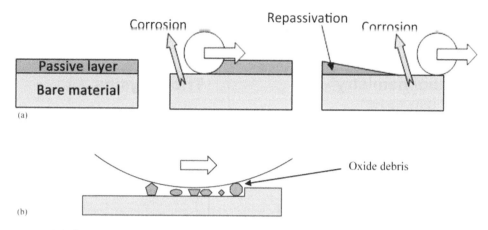

0.2 Synergistic effect of corrosion on wear and vice versa. (a) Corrosion accelerated by friction. (b) Abrasion accelerated by corrosion products

The calculation of the potential and current distributions in the contact can provide useful information on the risk of galvanic coupling between rubbed and not rubbed surfaces. It might also help to understand the degradation modes active on moving contacting surfaces. Research in this field is still in its infancy but noticeable progress is expected in the near future.

The synergistic effect of friction and corrosion is schematically represented in Figure 0.2. Corrosion can, for example, be accelerated by the mechanical removal of a surface film that causes bare material to be exposed to the electrolyte, leading to the onset of excessive corrosion of that bare material. After that, the corrosion of the bare material can be counteracted by a progressive re-growth of a passive surface film [4], a process known as 're-passivation'. On the other hand, solid corrosion products or debris resulting from the mechanical removal of the passive film may act as a 'third body' in the tribocontact [5]. Such debris consisting of hard particles such as oxides may induce abrasion on the bare material.

In aqueous environments, the composition of the electrolyte, the pH, and the content of dissolved oxygen are of primary importance. In the presence of water or humidity, the oxidation of tribological surfaces is known as 'tribocorrosion', and involves electrochemical processes such as corrosion and passivation [6]. A chemical surface modification may take place at the interface between a material and an aqueous solution due to adsorption processes or (electro)chemical reactions [7]. The concept of 'active wear track' has been proposed to describe depassivation and repassivation processes taking place in unidirectional sliding tracks immersed in an electrolyte [8].

The main fields of science to be taken into consideration when investigating tribocorrosion are summarised in Figure 0.3 and these concepts are clarified in Section 0.3

Tribocorrosion occurs in many different areas, from machinery (parts of engines, pumps, cutting tools, etc.) to medical engineering (hip and knee prostheses, fillings and caps, dental implants in physiological medium, etc.). Its study can provide some answers on the sustainability and ability of a material couple to function *in situ* in a mechanical system insofar as it is addressing in a multidisciplinary approach the

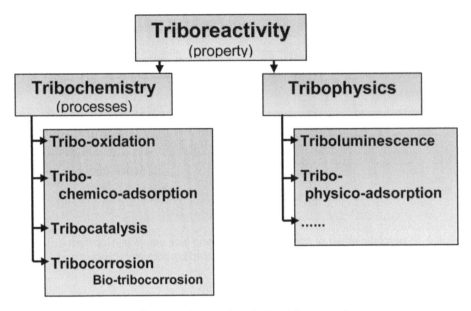

0.3 Main fields of science interacting during tribocorrosion process

trivalent system 'materials–environment–moving parts'. Since approximately 4–5% of the GNP of industrial countries is lost annually by corrosion and since 'tribocorrosion' may result in either an enhanced corrosion by wear and abrasion or an enhanced wear and abrasion by corrosion (Figure 0.4), it is expected that tribocorrosion may contribute up to 70% of that GNP loss.

Research and development work to enhance the knowledge of the phenomena causing tribocorrosion has thus a large industrial relevance. In addition, there is a strong need to develop appropriate testing procedures for tribocorrosion, and to develop models to understand and to minimise tribocorrosion effects. Tribocorrosion as a recently emerging field requires a scientific approach combining knowledge on corrosion and tribology, experimental investigation, and industrial validation.

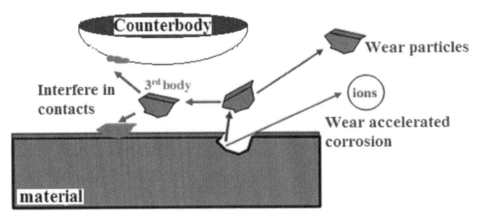

0.4 Schematic diagram of material degradation process during a tribocorrosion test

0.3   Definition of general concepts in triboreactivity

A proper definition of some concepts mentioned in Figure 0.4 is given hereafter for clarification to non-specialists.

Triboreactivity of a surface in contact with a surrounding medium (liquid or gas) is a property of that surface which results in its ability to become the seat of reactions (e.g. chemical, electrochemical, or adsorption), particularly when that surface is subjected to the action of mechanical loading (e.g. friction, abrasion, erosion, cavitation).

Tribochemistry refers to all reactions or chemical processes (chemical or electrochemical, and/or adsorption) taking place on surfaces subjected to a mechanical loading. Among tribochemical reactions and processes, we can mention tribo-oxidation, tribocorrosion, biotribocorrosion, and tribocatalysis. In this context, it is possible to give the following simple definitions of these concepts.

Tribo-oxidation is the oxidation process occurring on surfaces undergoing a mechanical impact in a reactive environment, typically oxygen in our environmental atmosphere. This process is usually activated by high temperatures and pressures imposed on the contact. This oxidation process affects the surface to a substantial thickness. By analogy, this concept can be extended to tribo-nitriding, tribo-sulphurising, etc.

Tribocorrosion is the degradation of material surfaces by the combined action of corrosion, electrochemical passivation and friction. It is essentially a surface process, but some events such as hydrogen evolution and absorption by the material, can cause a loss of mechanical properties of the subsurface material. By analogy, it is possible to extend this concept when, in the process previously described, a chemical or physical adsorption of inhibiting species strengthens or replaces the electrochemical passivation process. *Biotribocorrosion* (e.g. degradation of the moving parts in a knee prosthesis) is such a case. The successive repetition of some of these processes can lead to a possible synergy between mechanical stress and the effect of the environment that results in damage of surfaces and systems through an accelerated loss of functionality.

Tribocatalysis is the activation of a chemical or electrochemical reaction that takes place on a solid surface by the action of a mechanical loading in relative motion applied on a surface. This can be a way to maintain a high level of reaction kinetics over an extended time.

Triboluminescence is the physical phenomenon related to the property of certain crystalline materials to produce light under the action of rubbing, crushing or fracture. This phenomenon occurs in particular on materials or oxides doped with rare earth elements.

Adsorption is a physical or chemical phenomenon by which molecules in a liquid or a gas can attach themselves to the surface of a solid. The phenomenon of adsorption stems from the existence of unmatched forces that are physical in nature at the surface of a solid. In some cases, matched forces of a chemical nature, can also exist. The physical ones are electrostatic forces known as van der Waals forces. They are at the origin of a large number of adsorption phenomena, and they act rapidly and in a non-selective way, and do not induce any change to the adsorbed molecules. The process is reversible, and the force equilibrium varies depending on the surface of the solid, its porosity, and the temperature. The chemical ones involve chemical interactions between the molecule to be adsorbed and the adsorbent. In this case, adsorption is a selective process, leading to a modification of the adsorbed molecules, and is in general irreversible.

0.4   Specificity of laboratory and industrial tribocorrosion tests

As in classical mechanical testing, tribocorrosion tests can be grouped into two categories based on their different but complementary purposes, namely fundamental and technological tests.

Fundamental tests are implemented in research laboratories and their objective is to clearly identify and to understand, under well-defined testing conditions, the basic mechanisms and their synergy that govern the phenomena of tribocorrosion. These tests require the development of experimental methodologies for both the tests themselves and the techniques to be used for analysing and measuring data and other experimental outcomes. Concerning friction in particular, these tests are commonly performed on instrumented tribometers whose design and manufacturing are in line with the rules used, for example, for the manufacturing of traction machines. They must allow the application of controlled mechanical stresses and displacements on material couples exposed to air or aggressive liquid environments. The geometry of the samples is normalised as in mechanical testing. Two types of tests can be considered:

- tests at low displacement amplitude referred to as '*fretting tests*'. These tests provide information on the response of materials with respect to the solicitation (displacement amplitude, load, frequency, and environment). One can differentiate stick slip, partial slip or gross slip operating conditions. The information collected is on the nature of degradation, its location in the contact, the kinetics of crack initiation and crack growth, and the size and shape of the degradation products that may appear in the contact during testing
- tests at large displacement amplitudes, referred to as either '*reciprocating unidirectional sliding tests*' or '*continuous circular sliding tests*', provide information on the nature and kinetics of the wear process in connection with the synergies resulting from the mechanical, chemical or electrochemical coupling taking place on contacting surfaces in relative motion.

These tests allow the following analyses based on *in situ* and *ex situ* measurements:

- determination of the mean and local coefficient of friction
- identification of and study of the interactions between surfaces and environment, the nature of the mechanical-chemical coupling, the electrochemical or galvanic coupling due to a heterogeneous structure, and the shape and location of rubbed and non-rubbed areas
- establishment of local wear laws and their spatial distribution on the surface in view of a modelling of wear aiming at a future predictive approach.

The full investigation of the tribocorrosion tests generally requires the use of *in situ* and *ex situ* tools. Figure 0.5 gives an overview of the most commonly used additional tools in that respect in tribocorrosion testing.

Electrochemical techniques	Various "*ex situ*" techniques
- open circuit potential measurements	mainly, surface analysis techniques:
- polarization curves,	- optical microscopy,
- current transients,	- SEM,
- impedance spectroscopy,	- micro-topography,
- noise measurements, ...	- micro-hardness measurements,
used to study "*in situ*" the electrochemical surface behavior *like the mechanism and kinetics of corrosion and passivation*, and its dependence on testing time, and on tribological conditions *(load, speed, amplitude, frequency,...)*	- structural study techniques - chemical analysis techniques ... used to characterize the structure, topography, mechanical properties, chemical composition of surfaces, ..., and changes induced by tribocorrosion

0.5 Short list of *in situ* and *ex situ* measurement techniques

Technological tests are implemented in industrial laboratories and technical centres and these tests have to be as representative as possible of the industrial system through, for example, the nature of contact, the environment selected, the stress level applied, and the type of relative motion. They allow the validation of predictive models developed from the laws derived from fundamental tests. This requires the development of experimental methodologies related to the tests themselves but also related to the measurement techniques to be used. From an industrial point of view, a problem linked to tribocorrosion is solved when a combination of materials is found capable of withstanding the stresses and accommodating differences in relative motion between the two contacting surfaces. This has to be achieved for a defined physicochemical environment and a given type of contact.

A full control of the loads based on contact mechanics including thermal and tribochemical aspects, will allow, in the longer term, an unravelling of the localised load bearing by materials. This will pave the way towards the establishment of a generalised predictive modelling of combined wear and corrosion in mechanical contacts leading to an improved durability and performance of industrial systems.

0.5   Normalisation aspects related to (bio)tribocorrosion

The International Organization for Standardization (ISO) gives the following definition of a standard: 'Document established by consensus and approved by a recognised body that provides, for common and repeated use, rules, guidelines or characteristics for activities or their results ensuring an optimal level of order in a given context.'

A standard is a reference document on a given topic. It indicates the state-of-the-art on technology and know-how available at the time of writing. In order to be considered as a standard, the document must satisfy two conditions:

- the means and methods described should be reproducible on using and complying with the conditions as mentioned
- the document must have received full recognition.

Table 0.2   Short list of standards related to tribology

ASTM G 99-95a	Standard test method for wear testing with a pin-on-disc apparatus
ASTM G 133-95	Standard test method for linear reciprocating ball-on-flat sliding wear
ASTM G 77-93	Standard test method for ranking resistance of materials to sliding wear using block-on-ring wear test
ISO 20808	Fine ceramics (advanced ceramics, advanced technical ceramics) – Determination of friction and wear characteristics of monolithic ceramics by ball-on-disc method
DIN 50324	Tribology; testing of friction and wear model test for sliding friction of solids (ball-on-disc system)
DIN 51834-1	Testing of lubricants – Tribological test in the translatory oscillation apparatus – Part 1: general working principles
ASTM G 119-04	Standard guide for determining synergism between wear and corrosion

A standard is the result of a consensus reached by a process called normalisation. A standard is an indisputable common reference providing technical and commercial solutions. It is used to simplify contractual relationships. Normalisation constitutes a tremendous development tool for research and a vector of broadcasting for innovation. Some examples of standards used in the field of tribology are included in Table 0.2.

There are four types of standards:

- *Core standards*: they set the rules concerning terminology, acronyms, symbols, metrology (e.g. ISO 31: Quantities and units)
- *Specification standards*: they mention the characteristics, the performance thresholds for a product or a service (e.g. EN 2076-2: Aerospace series – Ingots and castings of aluminium alloys and magnesium – Technical Specification Part 2)
- *Standards for analysis and testing*: they mention the methods and means for conducting a test on a product (e.g. ISO 6506-1: Metallic materials – Brinell hardness test – Part 1: Test method)
- *Organisation standards*: they describe the functions and organisational relationships within an entity (e.g. ISO 9001: Management Systems – Requirements).

0.6   Interactions between testing and normalisation of tribocorrosion

Europe is challenged to strengthen its science and engineering in the field of tribocorrosion to maintain its competitive industrial position against Asia and the USA. Tribocorrosion is especially important in areas where gasoline or diesel fuels currently being used are changed to environmentally friendly fuels such as 'biodiesel', which introduce small amounts of water and other substances acting as electrolytes at the interfaces in the tribosystems and cause corrosive degradation of surfaces. In a few years, with increasing amounts of bio-products in gasoline, tribocorrosion will become a vital field of research. Tribocorrosion studies will also play a more prominent role in the characterisation of biomaterials (tribocorrosion in the human body, e.g. in prosthetics) and is expected to give new insights into wear or degradation mechanisms of coated materials under mechanical load, corrosion prevention and decorative thin coatings, etc. As an example, the lifetime of a fuel pump is reduced significantly when using biofuels.

The combination of fundamental knowledge of corrosion processes and tribology does not provide the necessary knowledge on tribocorrosion mechanisms at the nanometre and micrometre scale that can be exploited industrially. Automotive, ground transportation, and aircraft industries will be the most affected branches, but other fields will also suffer from tribocorrosion. As well as the positive influence of a better insight into tribocorrosion in human safety by reducing part damage and technical problems in transportation, there will be an important benefit on the saving of raw materials and fuel consumption.

Taking into account these considerations, in 2002, the Commission «Tribocorrosion» of CEFRACOR – France (a full member society of the European Federation of Corrosion) has established the EFC Working Group on TriboCorrosion (WP18). The necessity to innovate constitutes a major strategic axis of the Council of Europe that wants the European Union to become a global leader in knowledge, and to be both competitive and dynamic globally in the years to come. Conscious of these issues, European institutes of standardisation have initiated studies to see how normalisation can contribute to the innovation leap. One approach lies in merging the normalisation of research activities and industrial concerns to develop access to innovative solutions.

Upstream standardisation used as a tool of economic intelligence, allows us to capitalise on and broadcast inter-disciplinary knowledge. It allows us to identify evolutions required by actors of civil corporations (authorities, users, consumer), to anticipate future rules of the market, and even to influence their definition. A company that participates in the development of a standard while integrating it in its own innovation can thus facilitate access to the market for its products and increase its market share.

Downstream standardisation favours technology transfer on innovative products. Providing adequate terminology, methods of characterisation, measuring protocols of performance of procedures and products, it gives confidence to end-users, facilitates implementation, and thus creates the conditions favourable for acceptance and development of innovation.

In that context, since 2007, the EFC Working Group on Tribocorrosion (WP18) has participated in the project TRIBOSTAND under the umbrella ENIWEP-EUREKA (European Network for Industrial Wear Prevention – E! 3606) and promotes action towards the elaboration of a test protocol on biotribocorrosion and its validation. This protocol, related to the performance of tribological tests on passivating materials (stainless steels, titanium, aluminium) in aggressive environments, has been elaborated within a partnership between Ecole Centrale Paris (Lab. LGPM) and Katholieke Universiteit Leuven (Dept. MTM) in the field of 'Tribocorrosion'. This protocol aims at extending the applicability of the ASTM G 119-04 standard entitled 'Determining Synergism between Wear and Corrosion' towards passivating materials used under tribocorrosion conditions. This protocol takes into consideration that it has to be used by industrial companies, universities, and research centres wishing to perform sliding tests on passivating materials immersed in aggressive liquids.

0.7   Objectives and approaches developed in this book

0.7.1   Objectives of this book

The primary objective of this book is to address the '*why*' and '*how*' of performing tribocorrosion and biotribocorrosion tests, as well as the need for a well-described testing concept and protocol for passivating materials, and its pre-standardisation.

To that avail, this book provides:

– a description of the physical, chemical, and electrochemical background necessary for the correct application of experimental tools and techniques allowing a qualitative and/or quantitative analysis of tribocorrosion or biotribocorrosion tests
– how to solve complex problems by using structured and rigorous approaches including hypotheses and validations
– researchers, engineers, and technicians active at universities, research centres, and industrial companies with a good practice code of laboratory testing related to (bio)tribocorrosion
– a way to characterise the impact of tribocorrosion on the reliability and durability of tribological systems, and to prevent and to deal with risks
– a detailed description of a testing protocol for passivating metallic materials that allows an interlaboratory comparison of experimental data. This includes a detailed description of sample preparation before testing, the selection of appropriate testing conditions, the performance of tests, and the post-testing data handling.

0.7.2   Approaches developed in this book

The targeted audience consists of scientific and technical people active in:

– industrial laboratories in charge of R&D and quality control
– research laboratories investigating the mechanisms of tribocorrosion and biotribocorrosion
– technical centres in charge of services related to standard testing.

The authors have adjusted their academic, industrial, and experimental input keeping in mind the targeted audience. As a result, the different authors address in the following chapters complementary subjects in an easily understandable way. This allows the readers to acquire quickly the necessary background knowledge on tribocorrosion and biotribocorrosion if they are novices in the field. But at the same time, it provides experienced laboratory persons with the most advanced knowledge currently available in Europe in this multidisciplinary field.

The approach developed in this book is centred around the following subjects, namely:

– Phenomena of tribocorrosion in the medical and industrial sectors (Chapter 1).
– Depassivation and repassivation phenomena: Synergism in tribocorrosion (Chapter 2).
– Specific testing techniques in tribology: Laboratory techniques for evaluating friction, wear, and lubrication (Chapter 3).
– Specific testing techniques in tribology and corrosion: Electrochemical techniques for studying tribocorrosion processes *in situ* (Chapter 4).

- Design of a tribocorrosion experiment on passivating surfaces: Coupling tribology and corrosion (Chapter 5).
- Towards a standard test for the determination of synergism in tribocorrosion: Design of a protocol for passivating materials (Chapter 6).
- Towards a standard test for the determination of synergism in tribocorrosion: Detailed testing procedure for passivating materials (Chapter 7).
- Normative approach (Chapter 8).

References

1. J. Jiang and M. M. Stack: *Wear*, 2006, **261**, 954.
2. L. Benea, P. Ponthiaux and F. Wenger: *Wear*, 2004, **256**, 948.
3. M. Stemp, S. Mischler and D. Landolt: *Wear*, 2003, **255**, 466.
4. P. Jemmely, S. Mischler and D. Landolt: *Wear*, 2000, **237**, 63.
5. D. Landolt, S. Mischler, M. Stemp and S. Barril: *Wear*, 2004, **256**, 517.
6. J. P. Celis, P. Ponthiaux and F. Wenger: *Wear*, 2006, **261**, 939.
7. S. Mischler, A. Spiegel and D. Landolt: *Wear*, 1999, **225–229**, 1078.
8. I. Garcia, D. Drees and J. P. Celis: *Wear*, 2001, **249**, 452.

1

Phenomena of tribocorrosion in medical and industrial sectors

Jamal Takadoum

ENSMM 26 Chemin de l'Epitaphe, F-25030 Besançon, France
Jamal.Takadoum@ens2m.fr

Amaya Igartua

TEKNIKER Otaola 20. P.K. 44, SP-20600 EIBAR Gipuzko, Spain
aigartua@tekniker.es

Articulating systems are complex mechanical systems operating under a combination of sliding, rotation, slip, vibrations, weight bearing, and loading conditions. For example, in hip or knee joints and dental prostheses, the working parameters largely affect the lifetime of such contacting material systems. Articulating systems are present in engineering systems ranging from large scale ones such as in transportation, robotica, or offshore installations, down to small sized ones such as in microelectromechanical systems (MEMs). A commonly experienced problem in medical prostheses and in mechanical systems is the synergism between corrosion and wear occurring under field operation causing an enhanced material degradation. Achieving better performing articulating systems in clinical and engineering sectors is a real challenge in this early 21st century. In this chapter, an overview is given of the main parameters affecting tribocorrosion in medical and industrial systems, and this is illustrated with a few case studies.

1.1 Tribocorrosion: concept and definition

Corrosion and wear are two modes of degradation of materials which are well known and widely studied. While basic mechanisms of corrosion have been clearly identified, despite progress over the past 30 years by tribologists, fundamental processes responsible for friction and wear have not been fully identified and understood. Several fundamental questions remain unanswered today. Indeed, if the electrochemist is now able to describe the liquid/solid interface and knows how to calculate the dissolution of a metal immersed in a corrosive liquid, the tribologist still poorly understands the physical origin of friction and the elementary processes of wear. A better understanding of these fundamental phenomena may be made by theoretical physics (molecular dynamics) and nanotribology [1–4].

Given what has been said in Chapter 0, the study of material degradation by processes that involve corrosion and wear, is a difficult task. No serious fundamental approach can be attempted today with the tools and knowledge available. But given the importance of tribocorrosion in many applications, there is a strong need to develop specific models and devices to simulate and study phenomena involving wear and corrosion at the same time.

During tribocorrosion, the synergy between mechanical wear and (electro)chemical corrosion results in a total volume of removed material W_t which can differ from the sum of material removed separately by wear or corrosion. The volume W_t is given as a function of three components:

$$W_t = W^m + W^c + W^s \tag{1.1}$$

where W^m and W^c are the volume of material removed separately by the effects of wear and corrosion, respectively, and W^s represents the synergistic effect between wear and corrosion which can account for 20–70% of the total volume of material removed [5–8].

Equation 1.1 may also be written as follows:

$$W_t = W^m + W^c + W^{cm} + W^{mc} \tag{1.2}$$

with W^{cm} the effect of corrosion on wear, and W^{mc} the effect of wear on corrosion.

1.2   Main parameters affecting tribocorrosion

1.2.1   Techniques and set-up for tribocorrosion studies

Tribocorrosion studies can be conducted by coupling electrochemical methods to tribology testing. A typical set-up for tribocorrosion testing is shown in Figure 1.1 [9]. It consists of a polytetrafluoroethylene (PTFE) cell mounted on a ball-on-disc tribometer called a tribo-electrochemical cell. A loaded ceramic ball (the counter-body) slides against the specimen (the working electrode) while the tangential force is measured with a force sensor. The electrochemical potential of the specimen is controlled with respect to a reference electrode which can be a platinum wire.

Depending on the particular application, tests can be conducted using a number of different contact geometries (Figure 1.2).

1.2.2   Materials and environment

The degradation mechanisms of a material subjected to friction in a corrosive environment involve chemical and mechanical effects, known for a long time as the Rehbinder effect. Indeed, the surrounding environment (air, gas or liquid) inter-acts with a material surface, affects greatly the surface composition, and thus can consequently affect the tribocorrosion process.

Environment, mechanical properties, chemical composition and microstructure are the most important parameters that determine the material behaviour in the case of tribocorrosion. Under tribocorrosion conditions, the interaction between the surface of a material and the electrolyte may lead to the formation of a new compound whose properties differ from those of the bare material. This may lead to an enhanced or decreased surface resistance to friction, wear, and corrosion.

During friction, a third body whose formation and evolution depend strongly on contact stiffness and geometry in addition to vibrations [10], plays a crucial role in friction and wear. Wear debris may agglomerate to form a lubricating and a protective film or to act as abrasive particles [11, 12].

It is known in tribology that the microstructure of materials has to be carefully chosen. In the case of steels, phases such as ferrite, austenite and pearlite are relatively soft with a hardness ranging from 100 to 300 HV, and offer little resistance to wear. Hard phases such as martensite and bainite with a hardness ranging from 800 to 1000 HV are particularly well-suited when the surface is subjected to abrasive wear

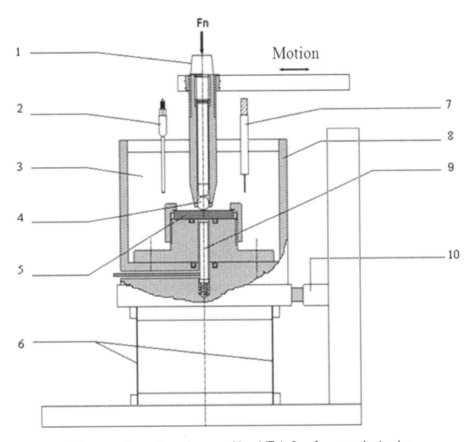

1.1 Tribocorrosion set-up: 1, normal load (Fn); 2, reference electrode; 3, electrolyte; 4, counterbody; 5, sample as working electrode; 6, spring plates; 7, counter electrode; 8, electrochemical cell; 9, connection of sample to potentiostat; 10, tangential force sensor (adapted from Ref. 9)

[13, 14]. In addition, the mechanical characteristics of steels can be greatly improved by the addition of chromium, molybdenum, vanadium, manganese and nickel. When added in small quantities (1–4%), these elements yield a finer microstructure and give the material greater hardness and wear resistance [14].

In tribocorrosion, things are more complicated since chemical reactions which take place at the interface between metal and electrolyte modify the composition, and consequently, the properties of the surface. Indeed, the surface of a hard material may oxidise and form brittle or soft phases as in the case of 34CrNiMo6 steel, titanium or tungsten. On the other hand, hydrogen embrittlement of some materials may take place, for example, in the case of iron and many types of hardened steels. Hydrogen reduction at the surface may also lead to the formation of brittle hydrides as in the case of titanium and zirconium [15].

1.3   Examples of tribocorrosion in medical applications

When orthopaedic or dental implants are placed in the body and submitted to friction, they may suffer tribocorrosion due to the simultaneous action of corrosion

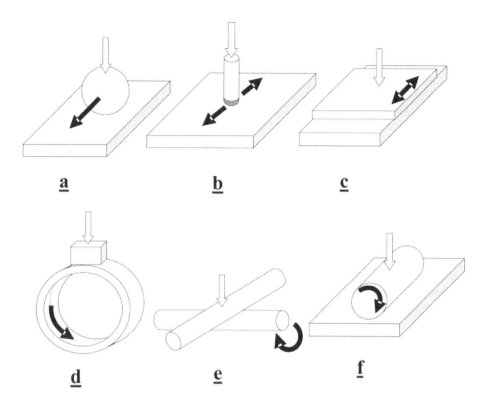

1.2 Some contact geometries used in tribocorrosion studies: a) ball-on-flat;
b) pin-on-disc; c) flat-on-flat; d) flat-on-cylinder; e) cylinder-on-cylinder;
f) cylinder-on-flat

by the body liquid and mechanical wear. This short overview is not intended as
a state-of-the-art and interested readers are suggested to consult specialised open
literature on biomaterials and their applications.

1.3.1 Biomaterials, biocompatibility

Biomaterials are natural or synthetic materials used in medical devices to perform,
restore, or replace the natural functions of living tissues or organs in the animal or
human body. The materials must be biocompatible which means that they will not be
rejected by the body and will not cause any infection. Due to their high passivity and
biocompatibility, 316L stainless steel, titanium and CoCrMo alloys are the metals
most used for medical implants (e.g. hip, knee, and tooth). Numerous polymers
such as ultra-high molecular weight polyethylene (UHMWPE) or ceramics such as
alumina or zirconia, are also widely used as biomaterials.

    One of the most important issues in the use of metallic biomaterials is their corro-
sion behaviour. The corrosion behaviour of an implant is influenced by a wide variety
of factors, including the material itself (chemical composition, microstructure,
surface condition), the surroundings (pH, temperature, O_2 content), as well as the
assembly itself (presence of crevices). The most common metallic biomaterials used
in implant manufacturing are compiled in Table 1.1:

- **Stainless steel**: In medicine, the stainless steel typically used (AISI 316L, ASTM F-55 and F-138) contains 17–20% Cr, 13–15% Ni, 2–3% Mo, and small amounts of other elements [16]. The notation 'L' indicates that the alloy contains a low amount of carbon (<0.03%) and is therefore not susceptible to intergranular corrosion that occurs due to the precipitation of Cr-carbides at the grain boundaries. Cr is the element mainly responsible for the high passivation ability of these alloys. An increase in Cr content, as well as in Mo content, strongly increases the resistance against a localised breakdown of passivity. Stainless steel implants are used as temporary implants to help bone healing, as well as fixed implants such as artificial joints. Typical temporary applications are plates, medullar nails, screws, pins, sutures and steel threads and networks used in fixing fractures. Use of steel in joint replacements has decreased since Co- and Ti-based materials became available. However, steel joints are still very popular and have an appreciable market share. The possible conversion of skin sensitivity to Ni or Cr by patients receiving stainless-steel-based implants has also restricted its use although there is no evidence to support this danger [17].
- **Co-based alloys**: As in stainless steels, the alloying of Co with Cr greatly enhances the corrosion resistance [18–20]. Since the passivation of pure Co takes place only in alkaline solutions, Cr is the key alloying element in Co alloys with regard to corrosion resistance in oxidising media. Mo has been found to be beneficial, especially under active corrosion conditions such as in hydrochloric acid. Co-based alloys used in total joint replacement surgery are cast CoCrMo alloy (Co28Cr6Mo, ASTM F-75) and wrought CoCrNi alloys (CoCrWNi, ASTM F-90 and Co–NiCrMo, ASTM F-562). The microstructure of the alloys and the resulting properties depend strongly on the production route (casting or forging), degree of cold-working, and heat treatment. The wrought CoNiCrMo alloys appear to have slightly better corrosion behaviour than the as-cast CoCrMo alloy [18]. Contemporary CoCr alloys are superior to stainless steel, both in fatigue and wear resistance, and are therefore preferred in total joint replacements, in both supportive and articulating locations.
- **Titanium alloys**: The excellent corrosion resistance of titanium alloys results from the formation of very stable, continuous, highly adherent, and protective oxide films on metal surfaces. Because titanium metal itself is highly reactive and has an extremely high affinity for oxygen, these beneficial surface oxide films form spontaneously and instantly when fresh metal surfaces are exposed to air and/or moisture. In fact, a damaged oxide film can generally form itself instantaneously if at least traces (that is, parts per million) of oxygen or water (moisture) are present in the environment. However, anhydrous conditions in the absence of a source of oxygen may result in titanium corrosion, because the protective film may not be regenerated if damaged. The nature, composition, and thickness of the protective surface oxides that form on titanium alloys depend on environmental conditions. In most aqueous environments, the oxide is typically TiO_2, but may consist of mixtures of other titanium oxides including TiO_2, Ti_2O_3, and TiO. Titanium (cp-Ti, ASTM F-67) and its alloys typically used in biomedical applications (Ti-6Al-4V, ASTM F-136-02a and Ti-6Al-7Nb, ASTM F-1295-05) can be considered as the most corrosion-resistant of the alloys described here. This is based on the very high stability of the TiO_2 passive film that forms spontaneously on the alloy surface [21].

Table 1.1 Overview of metallic materials used as biomaterials

Alloy family	Material designation	Common name	UNS designation	ASTM Standard	ISO Standard
Stainless steel	Fe18Cr14Ni2.5Mo	316 L Stainless Steel	S31673	ASTM F 138	ISO 5832-1
	Fe18Cr12.5Ni2.5Mo, Cast	316 L Stainless Steel	Unassigned	ASTM F 745	–
	Fe21Cr10Ni3.5Mn2.5Mo	REX 734	S31675	ASTM F 1586	ISO 5832-9
	Fe22Cr12.5Ni5Mn2.5Mo	XM-19	S20910	ASTM F 1314	–
	Fe23Mn21Cr1Mo1N	108	S29108	ASTM F 2229	–
Cobalt-based alloys	Co28Cr6Mo Casting Alloy	Cast CoCrMo	R30075	ASTM F 75	ISO 5832-4
	Co28Cr6Mo Wrought Alloy 1	Wrought CoCrMo, Alloy 1	R31537	ASTM F 1537	ISO 5832-12
	Co28Cr6Mo Wrought Alloy 2	Wrought CoCrMo, Alloy 2	R31538	ASTM F 1537	ISO 5832-12
	Co28Cr6Mo Wrought Alloy 3	Wrought CoCrMo, 'GADS'	R31539	ASTM F 1537	–
	Co20Cr15W10Ni1.5Mn	L-605	R30605	ASTM F 90	ISO 5832-5
	Co20Ni20Cr5Fe3.5Mo3.5W2Ti	Syncoben	R30563	ASTM F 563	ISO 5832-8
	Co19Cr17Ni14Fe7Mo1.5Mn	Grade 2 'Phynox'	R30008	ASTM F 1058	ISO 5832-7
	Co20Cr15Ni15Fe7Mo2Mn	Grade 1 'Elgiloy'	R30003	ASTM F 1058	ISO 5832-7
	Co35Ni20Cr10Mo	35N	R30035	ASTM F 562	ISO 5832-6
Ti alloys	Ti CP-1	CP-1 (Alpha)	R50250	ASTM F 67	ISO 5832-2
	Ti CP-2	CP-2 (Alpha)	R50400	ASTM F 67	ISO 5832-2
	Ti CP-3	CP-3 (Alpha)	R50550	ASTM F 67	ISO 5832-2
	Ti CP-4	CP-4 (Alpha)	R50700	ASTM F 67	ISO 5832-2
	Ti3Al2.5V	Ti-3Al-2.5V (Alpha/Beta)	R56320	ASTM F 2146	ISO 5832-6
	Ti5Al2.5Fe	Tikrutan (Alpha/Beta)	Unassigned	–	ISO 5832-10
	Ti6Al4V	Ti-6Al-4V (Alpha/Beta)	R56400	ASTM F 1472	ISO 5832-3
	Ti6Al4V, Cast	Ti-6Al-4V (Alpha/Beta)	R56406	ASTM F 1108	–
	Ti6Al4V, Eli	Ti-6Al-4V ELI (Alpha/Beta)	R56401	ASTM F 136	ISO 5832-3
	Ti6Al7Nb	Ti-6Al-7Nb (Alpha/Beta)	R56700	ASTM F 1295	ISO 5832-11
	Ti15Mo	Ti-15Mo (Metastable Beta)	R58150	ASTM F 2066	–
	Ti12Mo6Zr2Fe	'TMZF' (Metastable Beta)	R58120	ASTM F 1813	–
	Ti11.5Mo6Zr4.5Sn	'Beta 3' (Metastable Beta)	R58030	AMS-T-9046	–
	Ti15Mo5Zr3Al	Ti-15Mo-5Zr-3Al (Metastable Beta)	Unassigned	–	ISO 5832-14
	Ti13Nb13Zr	Ti-13Nb-13Zr (Metastable Beta)	R58130	ASTM F 1713	–
	Ti45Nb	Ti-45Nb (Metastable Beta)	R58450	ASTM B 348	–
	Ti35Nb7Zr5Ta	'TiOsteum' (Metastable Beta)	R58350	F-04.12.23	–
Special alloys	Ta, Unalloyed, Cast	Unalloyed Tantalum (Alpha)	R05200	ASTM F 560	–
	Zr-2.5Nb	Zircadyne 705	R60901	ASTM F 2384	–
	Ni-45Ti	Nitinol (Intermetallic)	N01555	ASTM F 2063	–

1.3.2 Orthopaedic implants

If one considers reconstructive implants, implemented to replace damaged joints, material selection and surface treatment play a crucial role in achieving a smooth running, and this has to be done by considering:

–   first the choice of a pair of rubbing materials that are biocompatible and have excellent mechanical and tribological properties. One seeks in particular to reduce or to eliminate the emission of wear debris that is the source of serious complications such as osteolysis or loosening, then

–   a control of the surface characteristics of materials and particularly the area of contact between the prosthesis and the bone which should enable bone growth and promote osteointegration. Indeed, for hip prostheses (Figure 1.3), high bond strength between the implant and the bone is required to ensure a good reliability of the prosthesis. The femoral stem is often made from titanium-based alloy (usually TA6V4 alloy) coated with 100–300 µm thick hydroxyapatite coating which is attached to a 50–100 µm thick porous titanium undercoating. In the case of a loss of adhesion, the micro-movements between bone and prosthesis may lead to a total detachment of the stem.

When stainless steel, titanium-based alloys and Co–Cr–Mo alloys are implanted as orthopaedic devices, they become covered with a protective passive film consisting mainly of metal oxides and hydroxides which prevent corrosion. However, in the presence of fretting or any tribological contact such as rolling or sliding, the protective films may be abraded and removed exposing bare metal to the body fluid. The bare metal may then undergo corrosion. Both effects affect the long-term durability of implants.

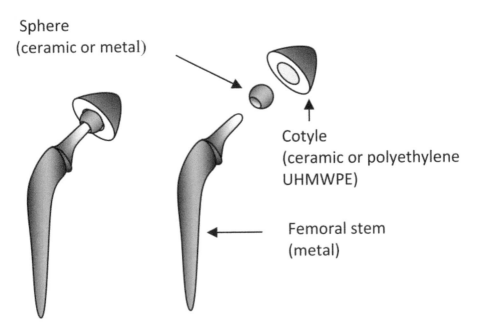

Sphere
(ceramic or metal)

Cotyle
(ceramic or polyethylene
UHMWPE)

Femoral stem
(metal)

1.3 Schematic diagram of the assembly of a hip joint prosthesis

Table 1.2   Main combinations of materials used for the femoral head and the acetabular oup in hip proothoooo

Femoral head	Acetabular cup
CoCrMo alloy	Ultra-high molecular weight polyethylene (UHMWPE)
Partially stabilised zirconia	UHMWPE
Alumina	UHMWPE
CoCrMo alloy	CoCrMo alloy
Alumina	Alumina

The main combinations of materials used in hip prosthesis for the femoral head and acetabular cup are shown in Table 1.2.

Hip prostheses constituted from a metallic femoral head in a polymeric (UHMWPE) acetabular cup (Charnley type) have dominated the market of hip joint replacement for 10 years [22]. However, UHMWPE wear leads to the formation of wear debris responsible for osteolysis and sometimes loosening of the implant. The amount and size of these debris are the key factors that determine both the occurrence and severity of osteolysis [23–25]. The use of alumina and zirconia femoral heads (instead of steel) articulating on UHMWPE acetabular cups allowed to reduce by half the wear of the polymer [24]. Nevertheless, more improvements were needed to ensure a significant increase in the life span of the prosthesis. This was the reason why there was recently an increasing interest on metal-on-metal and ceramic-on-ceramic implants. In these systems, both volume and size of wear debris are lower compared to metal-on-polyethylene ones. The wear rate of metal-on-metal hip prostheses was more than three orders of magnitude lower than for metal-on-polyethylene ones (1–6 mm^3 per year compared to 30–100 mm^3 per year). In addition, while polyethylene debris are found to be between 0.1 and 0.5 μm diameter, debris of only 0.02 μm diameter were detected in the case of metal-on-metal implants [26]. Particles in the nanometre size range were also reported [27]. However, due to their small size, metal debris disseminates easily in the body and may cause numerous diseases [28]. Consequently, a further lowering of material removal in metal-on-metal hip prostheses is still required. This is the object of numerous studies on biomaterials subjected to tribocorrosion.

Surface treatments and coatings are widely used to increase the life span of orthopaedic implants. Among the materials used for this application, ceramic coatings represent the most important class due to their good mechanical properties, their chemical inertness, and their low density. Diamond-like carbon (DLC) coatings were tested in a phosphate-buffered saline electrolyte at 37°C and pH 7.4 with three different surface finishes, namely as-deposited, polished with diamond and brushed [29]. The DLC coated samples with and without mechanical finishing had a lower corrosion activity under tribocorrosion conditions and also smaller wear tracks when compared to CoCrMo alloys. The corrosion of coated alloys was about two orders of magnitude lower on average and the coefficient of friction was only 50% of that recorded on uncoated alloy on sliding against alumina. A comparative study between the following four couples, namely 440C stainless steel ball against 440C stainless steel plate, Al_2O_3 ball against 440C stainless steel plate, ZrO_2 ball against 440C stainless steel plate, and DLC coated 440C stainless steel ball against 440C stainless steel plate, was done [30]. Tests were conducted during one hour in bovine blood serum.

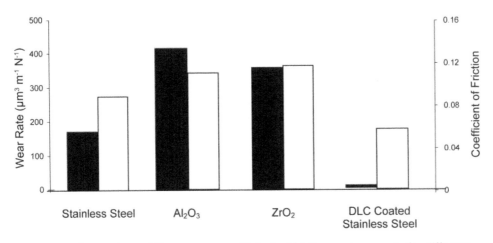

1.4 Comparison of the average coefficient of friction and wear rate for different material couples of interest as implant systems (adapted from Ref. 30). Solid bars, wear rate; open bars, coefficient of friction

The coefficient of friction and the wear rates are shown in Figure 1.4. DLC coated stainless steel ball against stainless steel plate clearly exhibits the lowest friction coefficient and the smallest wear. It is interesting to note here that DLC shows a better tribological behaviour than Al_2O_3 even though it has lower mechanical properties, namely a hardness of 18 GPa compared to 20.6 GPa, and an elastic modulus of 94 GPa compared to 391 GPa.

Corrosion and wear resistance of thermal sprayed coatings of nano ZrO_2 and Al_2O_3–13 wt%TiO_2 deposited on titanium were investigated [31] in Hank's solution. The alumina–titania coating showed a higher resistance to corrosion and wear than zirconia, but exhibits a higher coefficient of friction on sliding against alumina.

Further improvement of reliability and increase in life-span of orthopaedic implants require tribocorrosion studies with experimental apparatus representative of reality. Indeed, the results deduced from classical tribometers or triboelectrochemical cells are not fully representative of reality. Experiments with hip-joint simulators are needed to reproduce partially what really happens. Nevertheless, due to the difference in design and operating conditions (contact pressure, nature of motion, test duration, chemical composition of the experimental medium), it is often difficult to compare the results obtained. In addition, tests must be conducted in an electrolyte whose composition is as much as possible comparable to body fluid, instead of a saline solution as used in numerous corrosion/tribocorrosion studies of materials for implants. Indeed, the biological environment such as the presence of proteins, affects greatly both the corrosion and wear. Concerning biocorrosion, published results are often contradictory. Some publications suggest that the material removal increases in the presence of proteins while others argue the opposite [32–36].

1.3.3   Dental implants

During mastication, teeth are subjected to a tribocorrosion cycle since they are placed in a corrosive medium (saliva + acid or other chemical released by food) and subjected to friction (shear + impact). When dental implants or restorations are present in the mouth, they may suffer plastic deformation, wear and corrosion. In addition, due to the formation of electrolytic cells between stressed and unstressed areas, a metal ion release may lead to a thinning and consequently to a weakening of dental implants.

Titanium is one of the most frequently used materials in dentistry [37]. Corrosion [38] and tribocorrosion [39] of this material has been widely studied, generally in artificial saliva solutions. The addition of citric acid or an anodic inhibitor to artificial saliva results in a slight improvement in the tribocorrosion behaviour of Ti [39].

Dental implants made from titanium or TA6V4 alloy, consist of a metallic screw inserted into the bone. This screw is surmounted by another metallic or ceramic piece, the abutment, which provides a surface on which the artificial metallic or ceramic tooth can be placed (Figure 1.5). More sophisticated implants such as one piece dental implants made of zirconia have appeared in the market recently (Figure 1.6). The white colour of zirconia is an aesthetic advantage, its high strength to weight ratio, high corrosion resistance and high level of osteointegration and bone-to-implant contact explain the reasons and success of what is now called the 'new wave' dental implants.

The process of tribocorrosion of restorations or dental implants is in fact difficult to reproduce experimentally. Indeed, corrosion and wear processes are of several types (galvanic cell, stress corrosion, abrasion, erosion corrosion, plastic deformation and delamination). In addition, composition and pH of the liquid in the mouth vary continuously during eating and mastication. Sophisticated set-ups must be developed in future to reproduce the reality. The classical ball-on-flat tribometers do not fully reflect what really takes place in the mouth.

1.3.4 Chemical mechanical treatment in dentistry

In contrast to conventional dental caries treatment with rotary instruments and hand excavation, the chemical mechanical technique to remove caries is more selective since it allows removal of only carious material, it does not require anaesthesia, and it produces a dentine surface suitable for bonding with restorative materials. The technique consists of softening carious dentine by applying a gel for 30 s onto the

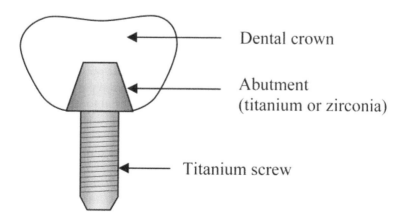

Dental crown

Abutment
(titanium or zirconia)

Titanium screw

1.5 Schematic diagram of a dental implant

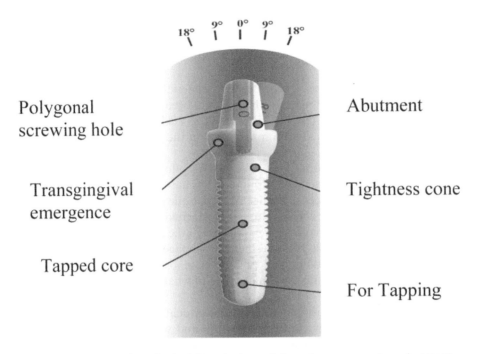

1.6 One piece zirconia dental implant consisting of a screw and an abutment (by courtesy of Paris Implants)

carious tissue, followed by a gentle excavation of the caries with specially designed instruments. This technique is minimally invasive, less painful for the patient, and in particular is recommended in paediatric dentistry as well for anxious or medically compromised patients [40–42].

1.4   Examples of tribocorrosion in industrial applications

1.4.1   Chemical mechanical polishing

Chemical mechanical polishing (CMP) is a particular tribocorrosion process very suited to prepare smooth and defect-free polished surfaces. It is widely used to polish hard and brittle materials using a solution containing an abrasive powder such as SiC, CeO_2, ZrO_2 or Al_2O_3. The pH of the solution and its chemical composition, the temperature, the hardness and concentration of the abrasive powder suspended in the solution, and the applied pressure are the main experimental parameters. They have to be carefully selected taking into account the base material to be treated. Chemical reactions which take place at the interface between solution and base material, generate reaction products at the surface which are softer than the base material. These products are subsequently removed by the abrasive action of the powder which must be softer than the base material to minimise mechanical damage to the surface. CMP is widely used in mechanics for the surface finishing of silicon nitride in bearing applications [43], and for the planarisation of microstructures grown on silicon wafers in microelectronics. CMP is also used in the silicon industry to fabricate semiconductor devices [44].

1.4.2 Industrial applications related to engines, cutting tools, pipelines, and nuclear reactors

Lubricants used in engines are expected to reduce friction and also to protect surfaces from corrosion. Indeed, corrosion inhibitors are added to oils to protect the moving engine components from corrosion by atmospheric oxygen or acidic products which form in the lubricants during operation. Some additives are used to neutralise acid formation whereas others form a protective film on the metal surface. During operation, the protective film may break up due to abrasive particles or oil ageing, and the bare metal may suffer from tribocorrosion by the joint action of acids and mechanical loads. This type of degradation leads to enhanced fuel consumption and exhaust emission whereas it decreases engine durability and reliability [45–47].

Cutting tools used for metal machining operated in the presence of lubricants or aqueous emulsions may also suffer from tribocorrosion in the same way as engines (see above). In the particular case of cutting tools for wood which operate without lubricant, it is worth pointing out that the cutting tool may suffer from tribocorrosion when the wood is an acidic variety like oak. In that case, the tool wear caused by tribocorrosion is larger than that noticed when the machined wood has a pH close to 7, for example, fir [48].

A joint effect of corrosion and wear is also encountered in the chemical and petrochemical industries, particularly in pipelines used for the transportation of corrosive liquids mixed with abrasive particles, such as in the case of acids containing ceramic particles or seawater or petrol containing suspended sand particles [49–51]. Indeed, in the petrochemical industry, numerous materials suffer from degradation resulting from the combined effect of sand and corrosive compounds present in the transported fluids.

The nuclear industry is also an industrial sector where tribocorrosion is encountered. Indeed, several metallic components used in pressurised water cooled nuclear reactors undergo erosion–corrosion and tribocorrosion such as the tubes and guides of the rod cluster control assemblies, and the gripper latch arms (GLAs) of control rod drive mechanisms [52–54]. In pressurised water reactors (PWR), the GLAs of the control rod command mechanisms (CRCMs) are protected from corrosion and wear by thick stellite 6 layers. Damage surveys on PWR have revealed that the wear, W, expressed as mass loss not only increases with the number, N, of mechanical interactions between a GLA and the control rod involving impact and sliding. It also increases for a given number of interactions with the average time interval, τ, between two successive steps, known as latency time. That wear is found to vary with the latency time as follows [55]:

$$W = N w_0 \left(1 + \frac{\tau}{t_0}\right)^{(1-n)} \qquad [1.3]$$

with n a dimensionless parameter whose value is close to 0.6, w_0 the average mass loss per interaction event, and t_0 a time constant. This time-dependent relationship was interpreted as the occurrence of a tribocorrosion process involving a combined effect of mechanical wear and corrosion in the following way. During successive interaction events, the satellite of a latch arm tooth is subjected to friction. However, metal loss by mechanical wear is not the prevailing consequence of friction. Stellite 6 is a chromium alloy that becomes passive in various corrosive aqueous environments including the PWR primary loop coolant. Friction induces a degradation of the passive film that protects the alloy. In the damaged areas, a dissolution of the alloy and a repassivation restoring the passive film are in competition. A metal loss occurs by dissolution of the damaged areas as long as the passive film is not completely restored. Equation 1.3 was first calculated from an empirical model [56] proposed to account for the passivation behaviour of stainless steel in high-temperature water. Based on lab tests [57], a wear law similar to the law for the GLAs (see Equation 1.1) was found. That law was explained based on a tribocorrosion mechanism involving a periodic mechanical depassivation followed, during the latency time, by a dissolution–repassivation process characterised by an anodic current, i, varying with time, t, as follows:

$$i = i_0 \left(1 + \frac{t}{t_0}\right)^{-n} \qquad [1.4]$$

1.5   Conclusions

Material degradation caused by tribocorrosion affects virtually all mechanical systems, for example, in mechanical engineering, the mining industry, oil refining, chemical synthesis, nuclear energy production, but also in transportation, civil engineering, and even in the health sector, in particular, artificial joints and prostheses. In the biological environment in particular, the presence of proteins affects the formation and stability of passivating films in an ambiguous way. Some publications suggest that material removal is increased in the presence of proteins but many others argue that the opposite occurs. This is certainly one of the most important subjects of research in the coming years. From the overview given in this chapter, in numerous studies devoted to articulating systems, the role of the surface oxidation of materials on the performance in moving contacts is clearly appearing. In all applications, the rate of formation, stability and mechanical properties of the passivating films play a crucial role in tribocorrosion phenomena. These passive layers may undergo friction forces acting in the contact areas. Their composition, thickness, structure, and mechanical properties are major parameters affecting the tribological behaviour of material surfaces. The next chapter will examine this aspect in detail. Today, modelling of the behaviour of materials operating under conditions where tribocorrosion is occurring is still in its infancy. Notwithstanding this, recent successful attempts to correlate field data with laboratory test results in the case of latch arm teeth in nuclear reactors are most promising.

References

1. G. M. McClelland and J. N. Glosli: in 'Fundamentals of friction: macroscopic and microscopic processes', (ed. I. L. Singer and H. M. Pollock), 405; 1992, Amsterdam, Kluwer Academic Publ.
2. B. Bhushan: in 'New directions in tribology', (ed. I. Hutchings), 141–158; 1997, London, Mechanical Engineering Publ. Ltd.
3. R. Bennewitz et al.: Adv. Eng. Mater., 2010, 12, (5), 362–367.
4. S. D. Kenny, D. Mulliah, C. F. Sanz-Navarro and R. Smith: Phil. Trans. Roy. Soc. A, 2005, 363, 1949–1959.
5. A. L. Grogan, V. H. Desai, F. C. Gray and S. L. Rice: Wear, 1992, 152, (2), 383–393.
6. K. Miyoshi: Surf. Coat. Technol., 1990, 43–44, (2), 799–821.
7. H. Moon-Hee and I. P. Su: Wear, 1991, 147, (1), 59–67.

8. T. C. Zhang, X. X. Jiang, S. Z. Li and X. C. Lu: *Corros. Sci.*, 1994, **36**, (12), 1953–1962.
9. J. Takadoum: *Corros. Sci.*, 1996, **38**, (4), 643–654.
10. S. Mischler: *Tribol. Int.*, 2008, **41**, (7), 573–583.
11. J. Takadoum, Z. Zsiga and C. Roques-Carmes: *Wear*, 1994, **174**, (1–2), 239–242.
12. M. Varenberg, G. Halperin and I. Etsion: *Wear*, 2002, **252**, (11–12), 902–910.
13. R. Leveque and M. Entringer: in 'Le livre de l'acier', (ed. G. Béranger, G. Henry and G. Sanz), 584–608; 1994, Paris, Publ. Tec et Doc, Lavoisier.
14. M. A. Moore: in 'Fundamentals of friction and wear of materials', (ed. R. A. Rigney), Vol. 73; 1981, Metals Park, OH, ASME.
15. J. Takadoum: 'Material and surface engineering in tribology', 103–105; 2008, London, ISTE, London.
16. M. F. Swiontkowski, J. Agel, J. Schwappach, P. McNair and M. Welch: *J. Orthop. Trauma*, 2001, **15**, 86–89.
17. D. F. Williams: in 'Biocompatibility of clinical materials', Vol. I, 99–123; 1981, Boca Raton, FL, CRC Press.
18. A. I. Asphahani: in 'ASM metals handbook', Vol. 13, Corrosion, 658–668; 1987, Metals Park, OH, ASM International.
19. P. Crook and W. L. Silence: in 'Uhligs corrosion handbook', 717–728; 2000, New York, Wiley Interscience.
20. R. Schenk: in 'Titanium in medicine', (ed. D. M. Brunette, P. Tengvall, M. Textor and P. Thoms), 145–170; 2001, Berlin, Springer-Verlag.
21. G. W. Stachowiak, A. W. Batchelor and G. B. Stachowiak: 'Experimental methods in tribology', Tribology Series, Vol. 44; 2004, Elsevier.
22. D. Dowson: *Proc. Inst. Mech. Eng. Part H, J. Eng. Med.*, 2001, **215**, (H4), 335–358.
23. A. Kobayashi *et al.*: Proc. of the 43rd Annual Meeting of the Orthopaedic Research Society, San Francisco, CA, 1997, 68, Orthopaedic Research Society.
24. J. L. Tipper *et al.*: in 'Friction, lubrication and wear of artificial joints', (ed. I. M. Hutchings), 7–28; 2003, Bury St Edmunds and London, Professional Engineering Publ.
25. S. C. Scholes, S. L. Smith, H. E. Ash and A. Unsworth: in 'Friction, lubrication and wear of artificial joints', (ed. I. M. Hutchings), 59–74; 2003, Bury St Edmunds and London, Professional Engineering Publ.
26. J. A. Williams and J. J. Kauzlarich: in 'Friction, lubrication and wear of artificial joints', (ed. I. M. Hutchings), 41–58; 2003, Bury St Edmunds and London, Professional Engineering Publ.
27. P. Doorn *et al.*: *J. Biomed. Mater. Res.*, 1998, **42**, 103–111.
28. R. M. Urban *et al.*: *J. Bone Joint Surg.*, 2000, **82**, (A), 457–477.
29. C. B. Santos, L. Haubold, H. Loleczek, M. Becker and M. Metzner: *Tribol. Int.*, 2010, **37**, 251–259.
30. B. Shi, O. O. Ajayi, G. Frenske, A. Erdemir and H. Liang: *Wear*, 2003, **255**, (7–12), 1015–1021.
31. C. Richard, C. Kowandy, J. Landoulsi, M. Geetha and H. Ramasawmy: *Int. J. Refract. Met. Hard Mater.*, 2010, **28**, 115–123.
32. G. C. F. Clark and D. F. Williams: *J. Biomed. Mater. Res.*, 1982, **16**, 126–134.
33. A. Kocijan, I. Milosev and B. Philar: *J. Mater. Sci. Mater. Med.*, 2004, **15**, 634–650.
34. R. L. Williams, S. A. Brown and K. Merritt: *Biomaterials*, 1988, **9**, 181–186.
35. J. J. Jacobs and R. M. Chicago: *J. Bone Joint Surg.*, 1988, **80A**, 268–282.
36. J. Zhu, N. Xu and C. Zhang: *Adv. Contracept.*, 1999, **15**, 179–190.
37. J. Lindigkeit: in 'Titanium and titanium alloys', (ed. C. Leyens and M. Peters), 453–467; 2003, Weinheim, Wiley-VCH Verlag.
38. J. C. M. Souza, R. M. Nascimento and A. E. Martinelli: *Surf. Coat. Technol.*, 2010, **205**, (3), 787–792.
39. A. C. Vieira, A. R. Ribeiro, L. A. Rocha and J. P. Celis: *Wear*, 2006, **261**, (9), 994–1001.
40. L. Fluckiger, T. Waltimo, H. Stich and A. Lussi: *J. Dent.*, 2005, **33**, (2), 87–90.

41. H. K. Yip, A. G. Stevenson and J. A. Beeley: *J. Dent.*, 1995, **23**, (4), 197–204.
42. J. A. Beely, H. K. Yip and A. G. Stevenson: *Br. Dent. J.*, 2000, **188**, (8), 427–430.
43. M. Jian, N. O. Wood and R. Komanduri: *Wear*, 1998, **220**, (44), 59–71.
44. G. J. Pietsch, Y. J. Chabal and G. S. Higashi: *Surf. Sci.*, 1995, **331–333**, 395–401.
45. S. C. Tung and M. L. McMillan: *Tribol. Int.*, 2004, **37**, (7), 517–536.
46. E. P. Becker: *Tribol. Int.*, 2004, **37**, (7), 569–575.
47. N. K. Myshkin, C. K. Kim and M. I. Petrokevets: in 'Introduction to tribology', 214; 1997, Seoul, Cheong Moon Gak.
48. M. Gauvent, E. Rocca, P. J. Meausoone and P. Brenot: *Wear*, 2006, **261**, (9), 1051–1055.
49. G. Wilkowski, D. Stephens, P. Krishnaswamy, B. Leis and D. Rudland: *Nucl. Eng. Des.*, 2000, **195**, (2), 149–169.
50. A. Neville and C. Wang: *Wear*, 2009, **267**, (11), 2018–2026.
51. X. M. Hu and A. Neville: *Wear*, 2009, **267**, (11), 2027–2032.
52. D. Kaczorowski and J. Ph. Vernot: *Tribol. Int.*, 2006, **39**, (10), 1286–1293.
53. D. Kaczorowski, P. Combrade, J. Ph. Vernot, A. Beaudouin and C. Crenn: *Tribol. Int.*, 2006, **39**, (12), 1503–1508.
54. G. Kermouche, A. L. Kaiser, P. Gilles and J. M. Bergheau: *Wear*, 2007, **263**, (7–12), 1551–1555.
55. E. Lemaire and M. Le Calvar: *Wear*, 2001, **249**, (5/6), 338–344.
56. F. P. Ford and P. L. Andresen: Proc. 3rd International Symp. on Environmental Degradation of Materials in Nuclear Power Systems – Water Reactors, Traverse City, MI, 1987, 789–800.
57. L. Benea *et al.*: *Wear*, 2004, **256**, 948–953.

Depassivation and repassivation phenomena: synergism in tribocorrosion

Roland Oltra

Université de Bourgogne, I.C.B, UMR CNRS 5209, F-21078 Dijon Cedex, France

roland.oltra@u-bourgogne.fr

Tribocorrosion occurs under various modes leading in many conditions to a more or less continuous abrasion, i.e. scraping and removal of metal, as found in sliding contacts. The resulting active bare material may then undergo an electrochemical reaction with the fluid that results in a material loss. In between successive contact events, the bare material has the ability to repassivate partly or fully. In this chapter, the electrochemistry of passivating materials such as stainless steel or titanium alloys during tribocorrosion is considered.

2.1 Electrochemical properties of materials relevant for tribocorrosion

In this chapter, passivity is considered from a dynamic point of view without addressing the numerous studies devoted to structural and chemical composition. Attention is given to the kinetics of dissolution and film repair processes taking place after depassivation. Passive films are as thin as a few nanometres, and act as a barrier layer between metal and environment. Under steady-state conditions, the quality of the passive film is characterised by the ion and electron transport properties that determine the dissolution rate of passive metals. The reader will find detailed information on the passivity of materials in the references listed under Ref. 1.

2.1.1 Properties and electrochemical characteristics of materials in a passive state, influence of the solution chemistry

Passivity of alloys is a phenomenon of major technical importance that permits protection of materials from general corrosion. Nevertheless, passive materials are sensitive to localised corrosion such as pitting corrosion, crevice corrosion, or stress corrosion cracking. Under tribocorrosion conditions, the integrity of the passive surface film can be affected by mechanical loading but can also be damaged by a local dissolution of the bare material. Basic knowledge on passivity is therefore important, and is detailed in this chapter. The two most important parameters when modelling is undertaken, are on the one hand, the rate of repassivation from which a critical time for repassivation can be derived. This defines the time of exposure of the bare material surface to the aggressive electrolyte. On the other hand, the rate of dissolution of the bare material surface is a determining parameter.

Chemical composition of passive films is probably the most studied characteristic of passive films. *Ex situ* surface analytical techniques such as X-ray photoelectron spectroscopy (XPS) have been largely used to demonstrate that passive films are multilayered systems consisting of bilayers, namely an inner and an outer layer. The composition of such films depends on various parameters such as potential, electrolyte composition, and/or temperature. One important parameter is the ageing of passive films which is of practical importance in the case of a long-term evaluation of material resistance. The structure of such thin passive films has recently been studied by *in situ* techniques to avoid any doubt about the effects resulting from the removal of test samples from the aqueous phase in which they were formed. These investigations confirmed that, for different metals such as Ni and Cu, the passive film has a crystalline structure.

Passivity concerns non-noble metals for which environmental conditions (electrolyte) favour the dissolution or the formation of a second phase film which is usually a hydroxide or oxide film corresponding to the passive state as shown schematically in Figure 2.1.

On this basis, passivation can be investigated by electrochemical measurements (Figure 2.2) from which the different steps associated with this reaction scheme, can be derived. A critical potential for passivation (known as the Flade potential) exists at which a drastic drop of the anodic current (oxidation current) occurs. In the prepassive potential domain, adsorbates such as $M(OH)_{ad}$ block the active dissolution, and when the surface is completely covered with them and the deprotonisation is complete, the surface is covered by an oxide (MO) 3-D layer. As shown in Figure 2.2, in the absence of any passive film, the same metal can dissolve (see dashed line in Figure 2.2). This non-steady state dissolution is more difficult to control experimentally but the transient dissolution can be induced by a passive film breakdown due to a chemical of mechanical interaction. This transient phenomenon is defined as a depassivation–repassivation process, and its quantitative characterisation is the basis of any evaluation of material modifications due to tribocorrosion.

Thermodynamics of oxide formation are represented in pH–potential diagrams or Pourbaix diagrams [2]. The passivation potential is defined for bulk oxides. The kinetics of reactions involved in the growth of a passive surface film can be described by considering different processes occurring either at the inner passive film/metal interface, in the film itself, or at the passive film/electrolyte interface. A high field assisted model [3, 4] is frequently used to describe the growth of an oxide film. The dependence of the current, I, on the field strength, F_{ox} over the oxide, is given by:

$$I = A \exp (BF_{ox}) \qquad\qquad [2.1]$$

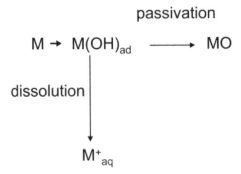

2.1 Reaction model for passivation with an initial stage of passivation of metal M explained by a 2D reaction model that involves adsorbed species and dissolution paths to M+ followed by 3D passive film growth

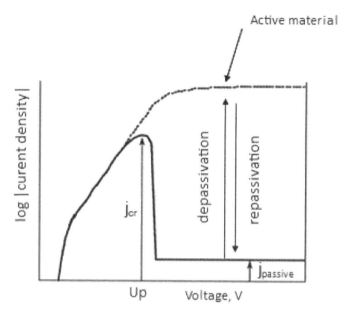

2.2 Typical anodic polarisation curve for a metal exhibiting passivation at U_p (Flade potential) (solid line). The dashed line represents the dissolution of the metal in the absence of passive film. Transient depassivation–repassivation processes represented at a fixed potential

with I the current density, A and B constants, and F_{ox} the field strength over the oxide. According to this model, the growth of the film follows an inverse logarithmic law as a function of time:

$$1/x = -B \log (t) \qquad\qquad [2.2]$$

with x the film thickness, and t the time. Considering cases of tribocorrosion, this approach will be of interest since it allows quantification of the rate of repassivation after mechanical damage (see Section 2.2.1).

Considering now the material degradation, the Faradaic balance, which is the evaluation of the species consumed or released through the passive layer during passivation or during a potential sweep in the passive range, is a useful parameter that can be monitored with a rotating ring-disc electrode (RRDE) [5, 6]. Such an electro-analytical technique provides quantitative information on the ratio of ionic species involved in the growth of a passive layer, and on the steady-state dissolution of a passive material. It allows the quantification of the rate of reactions during passivation. This electroanalytical technique also allows a follow-up of the mass balance in terms of chemical species during a passivation process (Figure 2.3) or during a transient response after a potential sweep in the passive range (Figure 2.4). In the case of a Fe–Cr alloy (Figure 2.3), the RRDE technique highlights the role of Cr species in the passivation process which stops the emission of Fe^+, i.e. the dissolution process observed in the active domain (see Figure 2.2). The second step described in Figure 2.1 corresponds to the oxide formation. Surface analyses are necessary to detect Cr(III) species that control the growth of the passive film [5].

The same technique can be applied to follow the non-steady-state behaviour of a passive film during a potential step. In such an experiment [6], the total current can

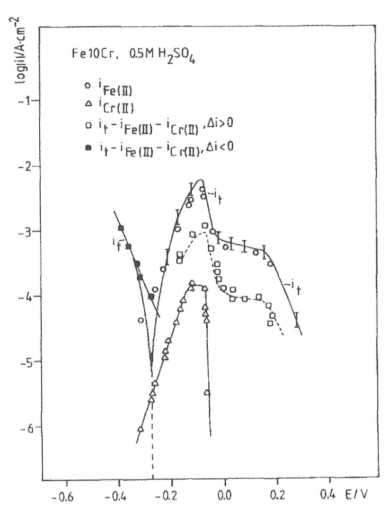

2.3 Stationary RRDE experiments on Fe–Cr during the quasi static $i_{Fe(II)}$ passivation in an acidic medium. i_t= total current (as described in Figure 2.2) and $i_{Fe(II)}$ and $i_{Cr(II)}$ the currents corresponding to the amount of Fe^{2+} and Cr^{2+} emitted, respectively [5]

be described as the sum of two contributions illustrated in Figure 2.1: the amount of charge devoted to dissolution and the amount of charge contributing to the film thickness change. This was demonstrated by RRDE, for example, during a potential step imposed on a passive material (Figure 2.4). The current collected on the ring after a potential step corresponds to the growth of the passive film and is the sum of these two contributions. In Figure 2.4, the collected current represents a reasonably high percentage of the current consumed by the dissolution versus the total current.

This approach can be compared to the chemical analysis of the electrolyte by atomic emission spectroelectrochemistry of electrolyte (AESEC). This technique is based on the use of an inductively coupled plasma optical emission spectrometer (ICP-OES). In this latter experiment, the analytical sampling of the electrolyte is placed downstream from an electrochemical flow-cell. This method was applied to

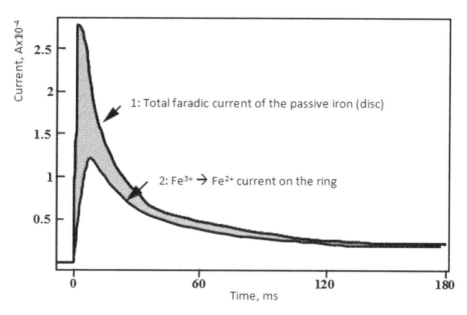

2.4 Charge balance after a potential step (0.4 V) in the passive range (no mechanical damage) measured with a RRDE (iron disc). Rate of dissolution of passive iron is controlled by the formation of trivalent Fe. Charge consumed during passive film growth is reasonably large compared to the charge related to the dissolution step [6]

follow the selective dissolution of Fe–Cr alloys during the passivation process [7]. Such a technique would be suitable to study transient phenomena but it has not yet been published.

An important feature of passive film reactivity is the electron transfer through the passive film. Such an approach can help to understand the galvanic coupling between depassivated wear track areas and the surrounding unaffected passive surface during tribocorrosion. Numerous studies have demonstrated that passive films behave like semiconductors even though their characteristics, defined mainly by photoelectro-chemistry or capacitance measurements, are very different from ideal semiconductor compounds.

2.1.2   Bare metal surfaces: how to reach a transient rate of dissolution?

The electrochemical phenomena on bare metal surfaces have been studied in many fields of electrochemistry, but as yet, not many attempts have been made to connect their outcomes.

Early works were performed using a 'triboelectrochemical' set-up consisting of a scratch test, to study the point of zero charge (pzc, briefly defined as the electrode potential where the charge in the electrical double layer is zero) of noble metals such as Ag, Au, and Pt [8]. In these pioneering studies, the objective was to reproduce the behaviour of the streaming Hg electrode largely used in fundamental electrochemical studies. The scrape potentials were controlled by: (i) the adsorption of various species from solution for systems in which the Faradaic currents are sufficiently slow, and (ii) the rates of various Faradaic reactions when the latter are rapid compared to the scraping process. The noble metals conform to case (i) over at least part of a pH range, and the scrape potential can be identified by the zero charge potential. This concept concerns the capacitive behaviour of the metal/electrolyte interface [9] but it has only been mentioned in a few papers related to tribocorrosion as the Faradaic processes (dissolution, repassivation) that balance the adsorption process. For example, the point of zero charge of a material is presented as a key parameter in sliding contacts [10]. However, that result was not obtained from the analysis of current or potential transients but from electrochemical impedance spectroscopy. From a phenomenological point of view, it is clear that the initial processes occurring after a mechanical loading generating a fresh metallic surface will be an adsorption and a rebuilding of the electrochemical double layer. This confirms the importance of knowing the charge balance during an electrochemical transient process after a mechanical impact.

The same trend was observed in electroanalytical techniques used to control the reaction kinetics at inert electrodes such as platinum or glassy carbon. The surface chemistry of such working electrodes contributes to the irreproducibility of kinetic measurements. In that field, there is thus a benefit to produce fresh and reproducible surfaces so as to achieve on solid surfaces results obtained on dropping mercury electrodes [11]. Various pre-treatments of solid electrodes were proposed such as mechanical polishing or chemical, electrochemical, thermal and pulsed laser activation. On glassy carbon electrodes, transient currents were found to depend on the electrode potential and the electrolyte solution. It was assumed that the linear dependence of the transient charges on potential implies that the *in situ* cleaning by laser is based on perturbation and restoration of the double layer (Figure 2.5). The inversion of the polarity of the response corresponds to the potential of zero charge of the glassy carbon in the tested solution. In that case, desorption of the double layer was

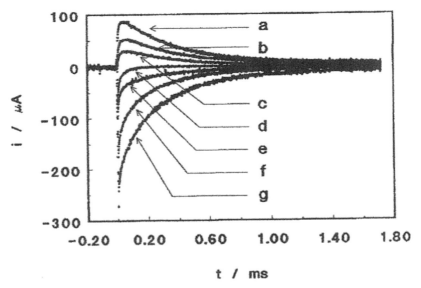

2.5 Laser induced current transients on glassy carbon electrode in 0.1 M KCl. Effect of the imposed potential varying from (a) –0.2 V, (b) –0.1 V, (c) –0.0 V, (d) 0.1 V, (e) 0.2 V, (f) 0.3 V, and (g) 0.4 V. Electrode was repolished before each transient [11]

assumed to be due to transient heating of the surface as the glassy carbon strongly absorbs the laser energy. The same trend was observed for passivating material when the oxide film was removed by a laser pulse [12]. This confirms that initial currents (peak current) observed on fresh metallic surfaces, whatever their origin, are associated with a charging of the double layer near the depassivated area [13].

This means that during mechanical breakdown, the balance between dissolution and repassivation charges will be a critical factor for the evaluation of the electrochemical dissolution in the damage area.

In situ electrochemical methods used in the field of tribocorrosion have been implemented up to now without asking the critical question about their validity even though a large amount of work has been performed to validate them [14]. Some interesting discussions can be found in a paper on mechanical abrasion [15]. Considering only electrochemical reactions, the question of the transient rate of dissolution of bare material surfaces is not yet clearly answered. In papers related to tribocorrosion, this latter point is rarely discussed. For example, the increase in the dissolution current in potentiostatic experiments resulting from rubbing on a passive surface is only characterised by recording the net current (Figure 2.6).

The increase of the net anodic current clearly shows that under tribocorrosion conditions, the corrosion rate of passive metals can increase by orders of magnitude compared to static situations. However, the qualitative estimation of the proportion of the current related to only dissolution cannot be easily extracted. The net current is the sum of various contributions such as the dissolution of the bare metal, the possible electrochemical reduction of species on this bare metal surface, and the current related to the repassivation of each elementary surface affected by the rubbing. In other terms, it appears that the conventional evaluation of the mass loss rate by Faraday's law can be complex in tribocorrosion experiments. On the other hand, the

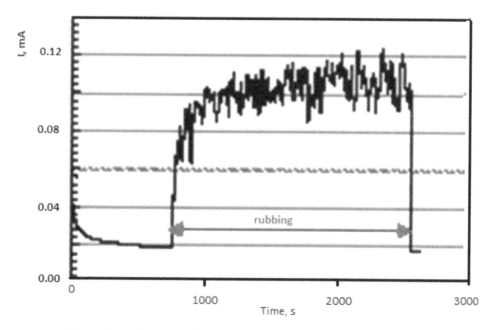

2.6 Evolution of the current recorded on Ti6Al4V rubbing against alumina in 0.9% NaCl at an imposed potential of 0.3 V vs. SSE [16]

current observed in Figure 2.6 cannot be considered as a continuous regime of depassivation (establishment of active dissolution in the worn areas) and repassivation of the activated areas. This latter point will be discussed in Sectioon 2.2.1.

The electrochemical analysis of the depassivation regime in tribocorrosion, whatever the contact mode might be, is limited by the rather limited actual knowledge of the elementary mechanical sequences occurring at the local scale. This can only be solved by simplifying the depassivation/repassivation sequence to take into account, for example, a third-body particle indentation on the passive material or a slurry impact in impinging jet experiments.

Potential sweeps from active to passive range show the importance of the control of the physical and chemical parameters to define carefully the electrochemical kinetics during active–passive transition. Such parameters are a nearly uniform current density over the bare metal surface, and well-defined mass transport conditions. Unfortunately, the most relevant electrode set-up that has been proposed, namely the shielded electrode developed in the 1920–1930 period [17], is not suitable for the analysis of tribological effects since the electrode is confined at the bottom of a cavity and no fresh surface can be generated. That kind of electrode is also defined as a recessed electrode and is largely used in the electroanalytical field.

A more common way is to study the electrochemical behaviour of microelectrodes with sizes in the range of tenths of micrometres, that allows a mimic of the behaviour of a mechanically depassivated area. On flat disc electrodes, it was confirmed that during the application of a potential step, the initial current peak is controlled by the ohmic drop [18]. These models were validated experimentally (disc electrodes at the bottom of a cavity) (Figure 2.7). All transients exhibit an initial current plateau the height of which is controlled by the ohmic resistance [19]. This is evidenced by the maximum current versus potential plot that shows a linear behaviour between the value of the current plateau and the applied potential. After an initial period of ohmic control corresponding to the plateau, the current decreases because of the growth of an anodic film. The most important part of the current transient in terms of charge, i.e. dissolution of the metal, is related to the plateau value. It can be

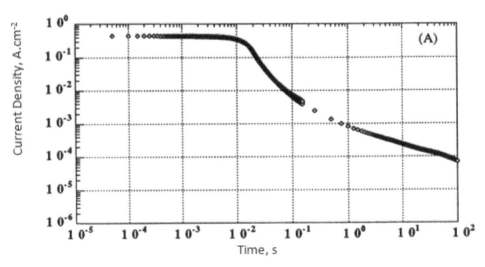

2.7 Passivation transients measured on AISI 430 steel disc electrode in 0.5 M H_2SO_4 at a potential of 0.5 V vs. SSE [19]

assumed that this feature (ohmic drop control) observed on such a recessed electrode could take place in tribocorrosion experiments.

This important feature has also been discussed in detail for mechanically disturbed electrodes on the basis of theoretical models considering the film growth kinetics and the ohmic drops in the electrolyte between the wear scar and reference electrode. But the agreement of the current calculated from these models with the experimental data obtained for tribocorrosion experiments, using an alternating motion tribometer, was only fair. The reason is probably that the mechanical parameters used in the models were not sufficiently representative of real situations. Nevertheless, it confirms that in electrochemical measurements performed under tribocorrosion conditions, the potential drop due to ohmic resistance is of major importance. Its value depends not only on the electrolyte conductivity, but also on the geometry of the contact which is more difficult to estimate as a function of the contact mode.

2.1.3 Electrochemical studies of the breakdown of passivity

The breakdown of passivity and consequently dissolution of bare metal appear in various types of localised corrosion but at different time and space scales:

- *Crevice corrosion:* chemical breakdown of passivity is induced by a change in chemistry of the electrolyte in contact with the passive film. Oxygen depletion occurs because of a limited diffusion from outside the crevice, and constitutes the initial stage necessary for the evolution of the crevice environment by local suppression of the cathodic reaction. Due to transport control, there is an increase in the concentration of metal cations and subsequently hydrogen ions produced by a metal–ion hydrolysis. The net anodic reaction inside the crevice creates an ionic current, with cations moving out of the crevice, and anions moving in. This leads to an acid solution inside the crevice where there is a progressive degradation leading to a total disappearance of passivity. This phenomenon is characterised by its duration (time of incubation) which varies from a few hours to several years, and leads to the initiation of corrosion.
- *Pitting corrosion*: chemical breakdown is described in the literature through complementary or consecutive mechanisms as the adsorption of aggressive anions, a penetration mechanism, and/or a passive film breakdown [20]. This leads to stabilisation of a pit described by a metastable or stable galvanic coupling between a bare metal surface (the pit bottom) and the surrounding passive surface. The critical size of a stable pit is in the range of a few micrometres, and the timescale in the range of seconds.
- *Stress corrosion cracking* (or fatigue corrosion): one of the possible mechanisms proposed to explain stress corrosion cracking propagation, namely the slip-dissolution model, is based on an extremely localised anodic dissolution due to mechanical fracture of the passive film. Crack growth proceeds by a cyclic process of film rupture, dissolution and film repair. Basic studies of the film rupture model have been proposed (see Section 2.2.2). In the stress corrosion field, an interesting mechano-electrochemical technique, namely the scratch test, was largely developed in the past to quantify the depassivation–repassivation sequence [21].

The timescale of the electrochemical transients is controlled by ohmic and capacitive processes in a similar way to the transient responses in the case of a potential step.

The electrochemical response related to an individual or isolated mechanical impact, whatever the nature of the contact mode (slurry, third-body particle, etc.), can be schematically represented as follows (see Figure 2.8). Consider a transient coupling simulated by the switch S of the impedance of an impacted surface and the surrounding passive surface. The peak galvanic current flowing through the surrounding passive surface to the bare surface will be controlled by the electrolyte resistance of the impacted surface. On the other hand, the current decay during 'repassivation' is controlled by the resistance and capacitance of the unaffected surface. Under potentiostatic conditions, calculations of the current decay versus time showed that the time constant of the current transient depends on Re and C, the electrolyte resistance and the capacitance of the electrode, respectively [22].

From this simplified description, it can be established that the total charge for the repassivation reaction, Q_w, originates from various contributions, namely:

$$Q_w = Q_{cd} + Q_d + Q_f - Q_{cath} \qquad [2.3]$$

with Q_{cd} related to non-Faradaic reactions mainly the charging of the electric double layer, and with contributions related to Faradaic reactions, namely Q_d, the dissolution of the material, Q_f the formation of the passive film, and Q_{cath} the cathodic reaction on the depassivated area. This is a qualitative description of the processes encountered during tribocorrosion experiments. However, knowledge of the polarisation curve of a bare metal surface which would give an intrinsic law remains unresolved even though some experimental efforts, as described in the next section, have been proposed in the literature.

2.2   Electrochemical response of mechanically damaged surface films

2.2.1   Mechanical film-rupture concept

The mechanical film rupture model was initially described for the generation of individual film ruptures characterised by the same electrochemical transient (exponential

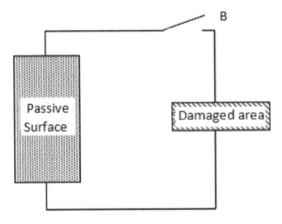

2.8 Equivalent scheme describing the role of local impedances of passive and mechanically damaged areas. For a potentiostatic device, the response to a mechanical removal of the passive film is controlled by the fast connection B of the two impedances

decay law) [23]. This mathematical concept was proposed to model the straining electrode test performed to evaluate the effect of straining on the dissolution of materials. The idea was to use the plateau value and the time constant of the current decay at the end of the straining to define the time constant of the current decay (Figure 2.9). The comparison between straining and mechanical impact of solid debris [24] demonstrated that the film rupture model can be applied for a multi-impact process.

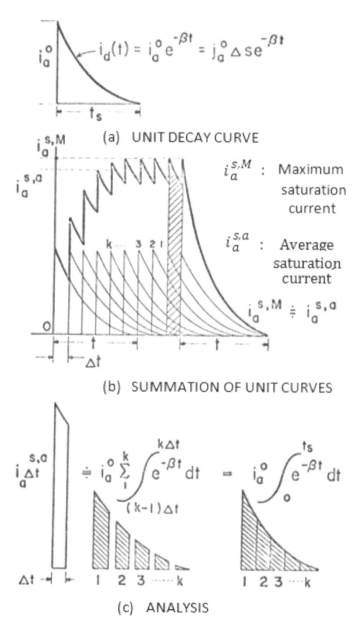

2.9 Description of the straining electrode theory applicable on any kind of cumulative electrochemical event, e.g. a mechanical removal of the passive film during rubbing [23]

In the case of tribocorrosion, whatever the contact mode is, the electrochemical responses due to wear particles that impact on the tested material can be represented using the same approach. However, in a real case, many physical parameters cannot be controlled and cannot be defined since they vary during the test. These parameters include the number of contacts per unit of time and space, and the size of the damage. This hinders the application of the 'straining electrode theory' because individual events cannot be averaged.

Some academic works have analysed ways to improve the time analysis of such stochastic processes. Acoustic emission (AE) coupled to electrochemical measurements during mechanical abrasion of a passive material have been analysed by spectral analysis confirming that the depassivation–repassivation events are not independent [25]. For example, erosion experiments in a corrosive environment were carried out on 316L SS instrumented with AE sensors. During the test, the sample was under potentiostatic control at a passive potential (+0.2 V $vs.$ saturated sulphate electrode) in 1 M H_2SO_4. The main problem to obtain accurate data was the validity of the charge measurements to define the mass loss caused by corrosion, which seemed to depend on the erosion rate. As shown in Table 2.1, under low abrasion conditions, the corrosion rate can be estimated by applying Faraday's law. In contrast, under high abrasion conditions, a discrepancy was always found between the corrosion rate estimated from Faraday's law, and the mass loss measured after the experiments corrected for the mass loss resulting from the mechanical erosion.

This discussion of the non-linearity of the electrochemical response is of importance to develop new models and to address the question of the validity of experiments based on a single mechanical damage. It should be noticed that this combination of electrochemical technique with acoustic emission, which could be easily transferred to other kinds of contact modes to evaluate the rate of mechanical erosion, has not yet been applied to our best knowledge, to cases of tribocorrosion (see also discussion on synergism in Section 2.3).

2.2.2 Simulation of mechanical breakdown (and recovery): conventional techniques and innovative approaches

The electrochemical behaviour of freshly generated metal surfaces on which very high reaction rates are expected, implies that the electrochemical measurements must be

Table 2.1 Experimental results for slurry erosion–corrosion experiments done under different abrasion conditions. Influence of the erosion rate estimated by the concentration of SiC particles (g L^{-1}) upon the mean value of the signals. <A(t)>= acoustic emission signal. Abrasion conditions: Type 316L SS, 1 M H_2SO_4, Eimp = +0.2 V/SSE– SiC 500 µm

SiC, g L^{-1}	<A(t)>, mV	Mechanical wear from <A(t)>, mg h^{-1}	Total mass loss, mg h^{-1}	Corrosive mass loss, mg h^{-1}	Mass loss calculated from Faraday's Law, mg h^{-1}
3	7	2.20	2.32	0.12	0.10
4	13	3.60	3.80	0.20	0.20
5	12	3.27	3.50	0.23	0.18
15	70	23.50	31.30	7.80	1.18
25	130	43.40	59.60	16.20	2.34

resolved in space. The influence of the ratio between depassivated and unaffected surfaces of the working electrode is a key factor [26–28]. It must be monitored at the very early stage of reactions on bare metal surfaces. Many techniques can be used to elucidate the kinetics and mechanisms of such reactions.

Repassivation has been investigated after a mechanical breakdown of the passive film by either straining [23], scratching [29], guillotining [26], or thin-film breaking [30]. In these techniques, the relaxation process starting immediately after the mechanical destruction of the passive film, was investigated by chronoamperometry. All of the electrochemical reaction steps for repassivation were discussed only on the basis of transient repassivation currents. One of the most used techniques is the scratching of a passive electrode using a static or rotating disc electrode [31]. In this latter case, a bare metal surface can be achieved when the surface oxide is mechanically removed by scratching the exposed surface of a rotating disc electrode with a diamond stylus. If the depassivated surface is continuously re-generated on the rotating electrode, the technique can represent a multi-impact process or an alternative rubbing process. The measured current results from the convolution product of the fresh surface generated by the elementary current response.

On the other hand, tribocorrosion can be studied by analysing the elementary response to a single mechanical event. For example, a spherical particle [32] or an angular abrading particle [15] can individually impinge a metallic target to record the individual transient current response. The two experiments confirmed that the predicted currents deviated from experimental values at higher particle concentrations where interactions between particles before and after impact, are not negligible. In the case of angular particles, a large damage distribution exists since not only indentation of the passive surface occurs but also scraping leading to a plastic deformation and the formation of a 'chip'. From a mechanical point of view, the impact of fragmented particles or the impinging impact of particles suspended in a fluid may cause an indentation of the metallic surface. This appeared to be the case even though the total kinetic energy of the impacted particle was partially released as elastic energy, but a more important part is used for mechanical scarring and plastic deformation. Based on this phenomenological description, some information useful for understanding what happens under tribocorrosion can be obtained by performing micro-indentation tests on the passive electrode. In that way, rupture and repair of a passive film on iron immersed in a de-aerated pH 8.4 borate buffer solution were investigated. Moving a conical diamond micro-indenter at a load of 0.1 N downward to the electrode and driving upward from the electrode, a couple of anodic current peaks were observed (Figures 2.10 and 2.11).

Nevertheless, the same difficulty arises for translating the transient response into a charge balance as in the case of a slurry impact. It is not possible to separate the different contributions, namely the capacitance recharge, the dissolution, and the amount of charge consumed in the repair of the film [33].

Up to now, this approach has been limited to the investigation of the mechanical properties of passive film [34]. As mentioned in the previous section, the main limit of the analysis of the electrochemical response associated with mechanical damage induced by a solid particle impact (slurry or third-body particle) is to find out how to discriminate the dissolution component from the total anodic current. This can be done by coupling a mechanical depassivation with a collection of cations released during the dissolution step. Such a study was first applied in the dissolution of Ti during repassivation in 6 M HCl using a scraping rotating ring disc electrode (RRDE)

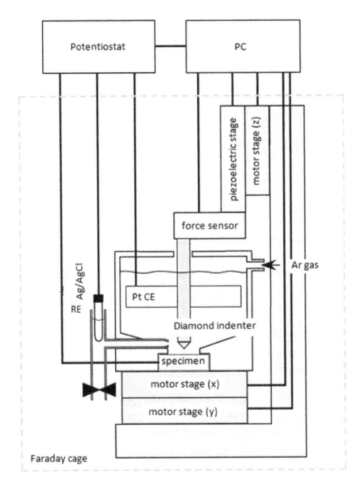

2.10 Schematic set-up for the *in situ* indentation of a passive metallic material [33]

[35]. The Ti^{3+} ions formed during scraping were detected on the ring electrode. Almost all of the disc current is consumed for the dissolution of Ti. The combination of mechanical depassivation with RRDE was also used for the study of repassivation of Fe and Fe–15 Mo binary alloys in an acetic acid solution [36]. The amount of iron dissolved during repassivation and its potential dependence were measured. The charge ratio between dissolution and total anodic reaction decreased on addition of 15% Mo. These promising approaches have not yet been applied to other systems because such experiments are not easy to perform.

To simplify the design of the electrochemical cell, a channel flow double electrode (CFDE) [37] was proposed to monitor the flux of metallic ions dissolved from the working electrode during repassivation after pulsed laser irradiation (Figure 2.12) [38]. The depassivation by laser irradiation is based on the mechanical decohesion of the protective oxide film induced by stresses developed at the inner film–metal interface [39].

The transient current during repassivation depends on the electrode potential and on the solution pH. The ferrous ions emitted were detected during repassivation with

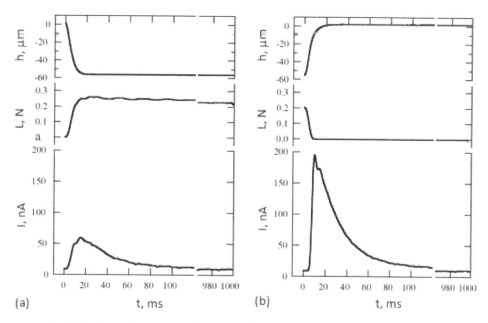

(a)   t, ms

(b)   t, ms

2.11 Study of elementary electrochemical responses during local indentation on a passive metallic surface. Time-transients of load (middle) and current (bottom) flowing through iron electrode polarised at 0.7 V vs. SHE in de-aerated pH 8.4 borate buffer solution: (a) downward and (b) upward movement of indenter at 10 mm s^{-1} at indentation depths, *h* (top) [33]

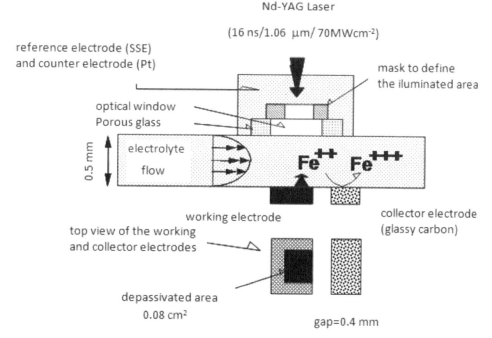

2.12 Channel flow double electrode (CFDE) combined with laser induced decohesion technique (LIDT) for a passive iron electrode. Details on calibration in Ref. 38

the CFDE, and the charge involved in the dissolution process, Q_d (see Equation 2.3), decreases with increasing electrode potential (Figure 2.13).

Secondly, the experimental responses were analysed assuming a reaction model with three dissolution paths (close to the diagram presented in Figure 2.1) to discuss in detail the contribution of dissolution and repassivation. The actual rate of repassivation was characterised in terms of the fractional coverage by trivalent passivating species and consequently only part of the total transient current is related to the dissolution of the metal (Figure 2.14). One of the limitations of the detection of emitted species by electrochemistry is the control of the rate of oxidation or reduction of the species on the collecting electrode. The same set-up was tentatively applied for tests on nickel and chromium but not on industrial alloys which would necessitate the simultaneous detection of the main alloying elements. Some attempts were performed by *in-situ* optical emission spectroscopy (laser induced breakdown spectroscopy) but calibration was difficult.

2.2.3   Galvanic effects

In the case of localised corrosion, galvanic current is assumed to occur between the 'cathodic' surface surrounding the localised corrosion site which behaves like the 'anode' as illustrated in Figure 2.15. The most efficient method to study such a current distribution is the scanning vibrating electrode technique (SVET) [40]. The vibrating probe records the current density profile above the electrodes (Figure 2.15).

The existence of a galvanic effect between the wear scar and the non-damaged (passive) surface is frequently mentioned in reports on tribocorrosion. However, it has not yet really been investigated at the local scale due to experimental limitations. During tribocorrosion, it is difficult to define the scale at which the galvanic process

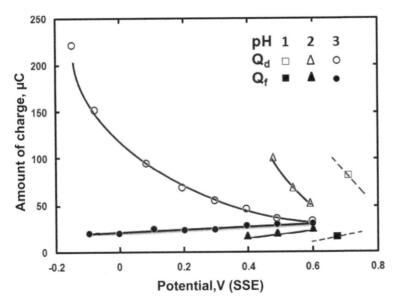

2.13 Potential dependence of charges involved in the dissolution of ferrous ions (Q_d) and in the formation of the passive film (Q_f) at three pH values [38]

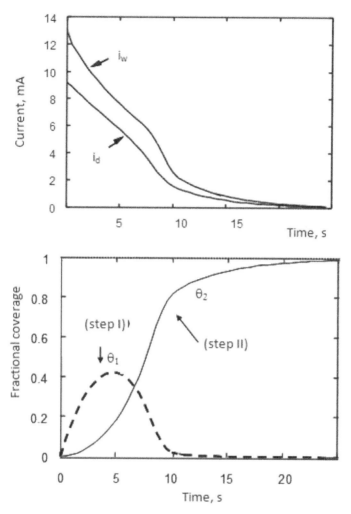

2.14 Transient current (dissolution current, I), and fractional coverages (θ_1 and θ_2) during repassivation of iron at 0.3 V/SSE in Na_2SO_4 solution pH 3. At $\theta_2=1$, the dissolution of Fe is terminated. θ_2 increases rapidly due to interactions between Fe(III) species. 3D growth of the passive film not taken into account in this model follows after this event [38]

takes place. Galvanic coupling can occur at the scale of the local mechanical damage (impact of abrasive debris and its surrounding) or at the scale of the wear track and the unaffected metallic surface. The measurement is impossible in the first case and not easy in the second one [41]. The problem might be that SVET is measuring the ohmic drop in front of the wear track induced by the contact of ball on a rotating electrode. The location of the anodic and cathodic areas is then not straightforward.

In unidirectional sliding tests (pin-on-disc), the access to the track and its surrounding is quite impossible. A bi-electrode set-up could be used to measure the galvanic current between the scratched surface and its surrounding surface (Figure 2.16) [42. This can be achieved by modifying the tested electrode which consists of an outer

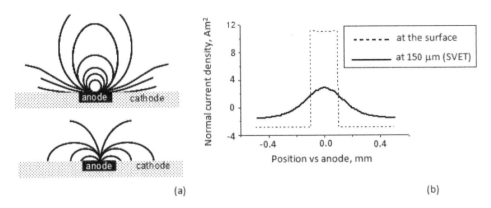

(a)   (b)

2.15 (a) Current distribution for anode (dissolving track 0.2 mm width) surrounded by cathode (passive surface). (b) Current density above the surface measured by scanning a SVET probe (dashed line) compared to real current density at the surface (continuous line, FEM solution of Laplace equation)

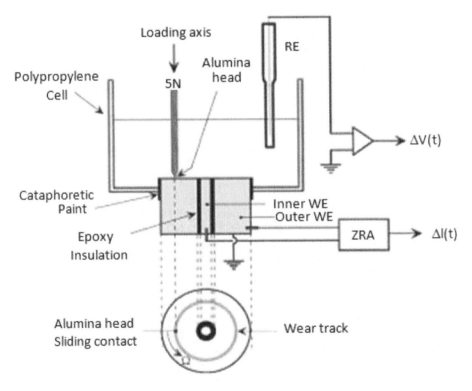

2.16 Electrochemical arrangement for monitoring tribocorrosion processes from current and potential fluctuations. Set-up derived from a RRDE allowing direct measurements of current flowing from wear track (ring) to unworn electrode (disc) [42]

electrode (Φ=25 mm) and an inner electrode (Φ=2.5 mm) electrically insulated. Only the outer electrode is continuously scratched by the pin, and the galvanic current is measured by means of a zero resistance ammeter (ZRA). With such a set-up, a galvanic current can be measured.

Other types of experiments have been proposed to measure the galvanic current in which the cathodic reaction is distributed between the undamaged surface of the tested surface and the surface of the counter electrode (CE). Even if this counter electrode is made of the same material, the experiment does not represent the real distribution of the galvanic current [43].

2.3   Synergism

Wear and corrosion involve mechanical and chemical mechanisms, and the combined action of these mechanisms often result in a significant mutual interaction.

2.3.1   Experimental evidence of interactions of mechanical and corrosion processes

The synergy between corrosion and wear processes is clearly exhibited by passive materials since, in the absence of mechanical removal of the passive layer, the rate of dissolution is reduced compared to the dissolution that is occurring, even under transient conditions, after mechanical damage. Nevertheless, the amount of material removed by chemical or electrochemical dissolution cannot be easily extracted from a total mass loss evaluation. The total mechanical mass loss can be estimated in some simple cases in the absence of corrosion (Figure 2.17). However, the difference between the total mass loss and the mechanical mass loss cannot be attributed to the chemical and electrochemical losses, as some enhancement of the mechanical damage can be observed in the presence of corrosion. This can, for example, be evaluated from the increase in third-body particles found in the wear scar [44].

The total damage, W_t, is expressed as the sum of the mechanical damage, W^m, and the electrochemical damage, W^c. The evaluation of synergy is only possible using simultaneous measurement of mass loss due to mechanical and chemical or electro-chemical dissolution. The mechanical removal on the target, W^m, can be easily defined in absence of corrosion, i.e. in a non-aggressive solution or under cathodic imposed potential if the mechanical properties of the surface layer of the metal are not affected by hydrogen embrittlement, and if the repassivation process is not altered later by previous cathodic polarisation. The value obtained can be compared to the pure mechanical loss obtained from a theoretical mechanical law (if available, as, for example, for sliding [45]). However, the value cannot be validated in the presence of important corrosion, confirming that, in some conditions, the electrochemical polarisation can affect the mechanical removal itself.

Some questions also arise concerning another aspect of the synergism, namely, the effect of the induced mechanical deformation on the rate of dissolution. In other words, whatever the nature of the contact, the mechanical properties of the impacted surface could be modified (plastic deformation, work hardening), and consequently the electrochemical response can be affected too. The effect of an elastic stress on the surface reactivity of a metal in contact with an electrolyte can be evaluated using a thermodynamic approach of the exchange current and the equilibrium potential as a function of the applied or residual stress, i.e. far from corrosion processes occurring

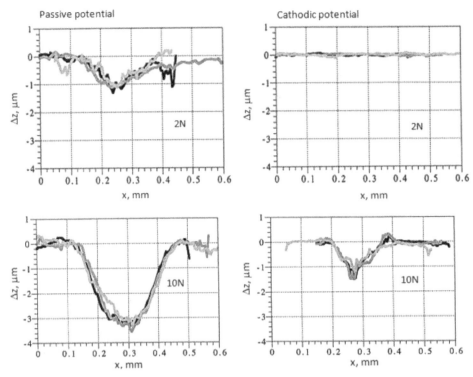

2.17 Cross section profiles of wear tracks after sliding tests at different normal loads and applied passive or cathodic potentials. These profiles illustrate the synergism between mechanical damage estimated from cathodic polarisation experiment and the material removed at passive potentials [44]

in tribocorrosion [46]. The main effect is always a decrease in the standard potential corresponding to a larger activation of the surface. On the other hand, tensile stress has been found to increase the current dissolution whereas the cathodic current would not be affected. But in the presence of plastic deformation as in tribocorrosion, all induced structural defects, e.g. dislocations, contribute to the increase in the dissolution current. The equation proposed to quantify this effect [46] is:

$$v = v_o (k\varepsilon + 1) \exp (\sigma_m V_m/RT) \qquad [2.4]$$

with v the corrosion rate of plastically deformed metal, v_o the corrosion rate of non-deformed metal, k a proportionality factor, ε the engineering linear strain, σ_m the absolute value of the hydrostatic component of the stress tensor, and V_m the molar volume.

Nevertheless, a clear description of the mechano-electrochemical theory is not really possible yet in the case of tribocorrosion since the local current from the track remains difficult to evaluate and correlate with the local evaluation of mechanical stresses. These mechanical-electrochemical interactions are often claimed as being important controlling factors but no evident demonstrations of their 'weight' in tribocorrosion have yet been proposed.

The use of electrochemical methods to define the enhancement of dissolution due to mechanical effects has been discussed more specially for sliding contacts [47]. The conditions allowing the application of Faraday's law (see Equation 2.5) cannot be fulfilled if the cathodic reaction that occurs on the bare metal surface is not negligible or if the valence of oxidation can change as a function of the applied potential:

$$W^c = QM/nF\rho \qquad [2.5]$$

with W^c the volume of metal dissolved, $Q = \int I\, dt$ the total charge (integration of the measured current I over the duration of the experiment), M the atomic mass of the metal, n the charge number for the dissolution reaction (apparent valence), F Faraday's constant, and ρ the density of the metal.

Compared to homogeneous dissolution processes which can be evaluated by chemical analysis of the solution, even under dynamic conditions [7], the chemical analysis of the solution in real time or at the end the tribocorrosion experiment cannot be performed since the solid metal particles removed by impact can dissolve or even be directly analysed in plasma-based analytical techniques. This ends up in an overestimation of the dissolved material part. In a previous part of this chapter, it was mentioned that the difference between the total wear, W_t, and the pure mechanical wear, W^m, can be larger than W^c derived from Faraday's law [25, 42]. This discrepancy can be due to some errors in the determination of W^m or to uncollected current due to galvanic processes as suggested previously.

From the previous discussions, it appears that a continuous recording of the mechanical wear and the electrochemical response would clarify some aspects of the synergism in tribocorrosion. In mechanics, acoustic emission (AE) is used to monitor hard particle impact on metallic surfaces [48]. It was found that the integrated RMS AE signal gives a real-time estimation of the total mechanical wear during the abrasion experiments in corrosive media. This technique is based on the record of the elastic energy released by each mechanical impact. If this approach is valid for spherical particles, some critical points arise for angular particles (slurries, third-body particles, etc.) for which the elastic energy is not proportional to the mechanical damage as angular particles produce a mechanical removal. In other words, only in the presence of glass beads can the impact be compared to an indentation (Figure 2.18).

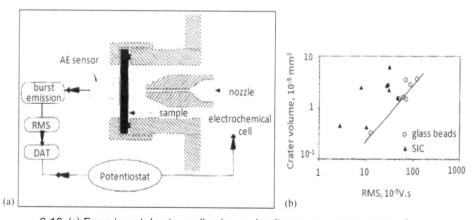

2.18 (a) Experimental set-up allowing a simultaneous measurement of mechanical and electrochemical damage by using acoustic emission (AE) for evaluating mechanical wear [48]. (b) Non-linear dependence of wear volume with impact rate for angular abrading SiC particles in contrast to linear correlation found for indentation by spherical particles [49]

Nevertheless, as the mechanical wear results from large cumulative individual processes, AE gives an average value that can be related to the average mechanical mass loss. A more detailed study on the correlation between the acoustic and electrochemical signals has been performed using electrochemical noise [25]. On plotting the power spectral density (PSD), a correlation was only found for a low regime of continuous slurry corrosion confirming that only at a low frequency of impacts can the current be considered as the superposition of elementary current transients.

2.3.2 Critical issues in the modelling of electrochemical-mechanical interactions

Several papers deal with deep modelling of the electrochemical response during tribocorrosion or abrasion corrosion [19, 50]. A theoretical approach was developed to take into account the film growth kinetics and ohmic drop in the electrolyte between the wear scar and reference electrode [19]. The electrolyte resistance associated with the scar was defined as follows:

$$R_\Omega = \frac{1}{2\pi\kappa b} \ln\left[2\left(1 + \frac{b^2}{a^2}\right)^{1/2} + \frac{2b}{a} \right]$$ [2.6]

with a the half-length, b the half-width of the electrode, and κ the conductivity of the solution. The model could simulate the general trends observed in current transients monitored during alternating motion tribometer experiments. Two models were developed:

– one considering **L**ateral **G**rowth (LG) of an oxide film on a mechanically depassivated surface
– one considering **U**niform oxide film **G**rowth (UG) as the rate determining process.

By combining these two models, it was possible to model the transient response with good accuracy (Figure 2.19), and to give a reasonably good qualitative description of the evolution of current with time (Figure 2.20).

An important quantitative discrepancy between simulated and measured current transients remains. This discrepancy can be attributed to mechanical and electrochemical factors not included in the model, such as the electrochemical conditions prevailing in the contact zone, the role of third-body particles, and the selective dissolution of iron during film growth. Nevertheless, an analysis of the model accuracy confirms that the repassivation kinetics of the alloy determines the chemical degradation rate in a tribocorrosion system rather than the intrinsic corrosion resistance. A second conclusion is that, in electrochemical experiments under tribocorrosion conditions, the potential drop due to ohmic resistance is of major importance. Its value depends not only on the electrolyte conductivity, but also on the geometry of the experimental arrangement, i.e. the sliding pin and the electrochemical cell.

Using interface kinetics as the growth limiting process, the current transients caused by the rubbing action for an inert pin-on-metal substrate (ipms) configuration and also for a metal pin-on-inert substrate (mpis) were modelled [50]. A solution was determined for the ipms configuration in a high conductivity electrolyte. One of the most important points was that the equations describing growth rate limited at a film interface are not only judged more appropriate for thin films from a physical aspect,

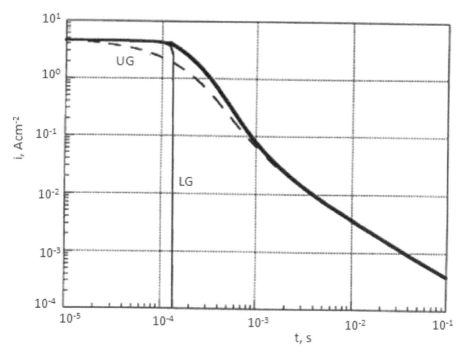

2.19 Combination of the (UG) and (LG) models describing limiting cases can be by assuming lateral growth up to monolayer coverage, followed by a growth in thickness [19]

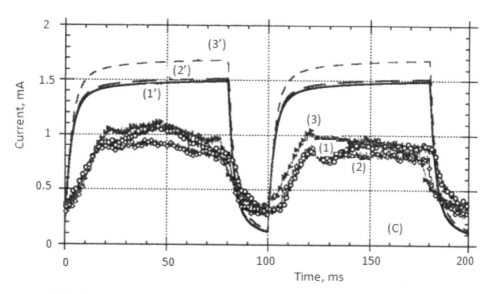

2.20 Calculated current transients (curves 1', 2', 3') for tribocorrosion experiments after 2, 10 and 30 min, respectively. The shape of the measured transients (curves 1, 2, 3) differs to some extent from the theoretically calculated ones. More importantly, the absolute values of the measured currents are significantly smaller than those predicted by the model. Frequency: 5 Hz. E_{appl}: –0.25 V *vs.* SSE – AISI 430 steel in 0.5 M sulphuric acid [19]

they are also more easily treated from the numerical point of view. For a reaction rate limited at an interface, there is no explicit need to consider film nucleation on a bare metal surface, since the reaction rate also remains finite for zero film thickness (Figure 2.21).

On the other hand, it must be mentioned that the passive film growth parameters were selected based on electrochemical quartz crystal microbalance experiments. The model is based on an analytical solution which seems to be valid for electrolytes with a negligible ohmic drop. This probably imposes a limit on the applicability of this interesting approach.

2.4   Concluding remarks

From the examples reviewed in this chapter, it can be concluded that the repassivation kinetics of an alloy undergoing tribological damage determines its corrosion damage rather than its intrinsic corrosion resistance defined under steady-state conditions in the absence of any mechanical perturbation.

The quantification of the non-steady-state dissolution during consecutive depassivation–repassivation sequences has not yet been modelled with sufficient accuracy. The main reason is that the ohmic drop which is the important controlling parameter depends not only on the electrolyte conductivity but also on the geometry of the tribocorrosion experiment.

Individual mechanical damage can be used to reach a reasonably good description of the Faradaic balance but the question of the cumulative individual damage still remains to be resolved.

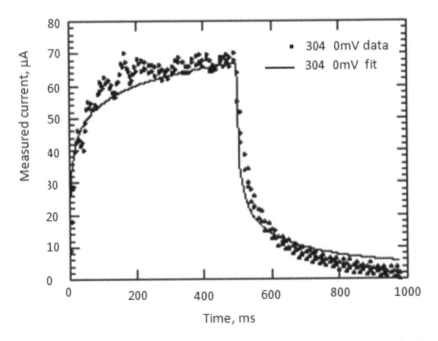

2.21 Current transient and fit for a 304PVD sample in sulphate solution [50]

References

1. References of proceedings of main passivity meetings:
   1a. R. P. Frankenthal and J. Kruger, ed.: 'The passivity of metals'; 1978, Princeton, NY, The Electrochemical Society.
   1b M. Froment, ed.: 'Passivity of metals and semiconductors'; 1983, Amsterdam, Elsevier.
   1c N. Sato and K. Hashimoto, ed.: 'Passivation of metals and semiconductors'; 1990, Oxford, Pergamon.
   1d K. E. Heusler, ed.: *Mater. Sci. Forum*, 1995, **185–188**.
   1e M. B. Ives, J. L. Luo and J. Rodda, ed.: 'Passivity of metals and semiconductors', Proc. Electrochemical Society, Vol. 99-42; 2000, Pennington, NY, The Electrochemical Society.
   1f P. Marcus and V. Maurice, ed.: 'Passivation of metals and semiconductors, and properties of thin oxide layers'; 2006, Amsterdam, Elsevier.
2. M. Pourbaix: 'Atlas of electrochemical equilibria in aqueous solutions'; 1974, Houston, NACE.
3. K. J. Vetter and F. Gorn: *Electrochim. Acta*, 1973, **18**, (4), 321–326.
4. R. Kirchheim: *Electrochim. Acta*, 1987, **32**, 1619–1629.
5. S. Haupt and H.-H. Strehblow: *Corr. Sci.*, 1995, **37**, (1), 43–54.
6. K. E. Heusler: *Ber. Bunsenges. Phys. Chem.*, 1968, **72**, (9–10), 1197–1205.
7. K. Ogle, M. Mokaddem and P. Volovitch: *Electrochim. Acta*, 2009, **55**, (3), 913–921.
8. T. N. Anderson, J. L. Anderson and H. Eyring: *J. Phys. Chem.*, 1969, **73**, (11), 3562–3570.
9. S. Trasatti: in 'Advances in electrochemistry and electrochemical engineering', (ed. H. Gerischer and C.W. Tobias), Vol. 10, 213–321; 1977, New York, Wiley.
10. J. Geringer, B. Normand, C. Alemany-Dumont and R. Diemiaszonek: *Tribol. Int.*, 2010, **43**, (11), 1991–1999.
11. R. K. Jaworski and R. L. McCreery: *J. Electrochem. Soc.*, 1993, **140**, 1360–1365.
12. G. M. Indrianjafy: 'Étude sur systèmes modèles des phénomènes transitoires de corrosion localisée des métaux passivables en solution. Perturbation de l'état passif par interaction laser-matériau', PhD thesis, Université de Dijon, France, 1993.
13. R. P. Wei and M. Gao: *Corrosion*, 1991, **47**, (12), 948–951.
14. S. Barril, S. Mischler and D. Landolt: *Wear*, 2004, **256**, (9–10), 963–972.
15. R. Oltra, B. Chapey and L. Renaud: *Wear*, 1995, **186–187**, 533–541.
16. S. Mischler: *Tribol. Int.*, 2008, **41**, 573–583.
17. J. W. Muller: *Trans. Faraday Soc.*, 1931, **27**, (Pt. 12), 737–751.
18. R. Oltra, C. Gabrielli, F. Huet and M. Keddam: *Electrochim. Acta*, 1986, **31**, 1501–1511.
19. P. Jemmely, S. Mischler and D. Landolt: *Wear*, 2000, **237**, 63–76.
20. H. Strehblow: in 'Corrosion mechanisms in theory and practice', (ed. P. Marcus), 201–237; 1995, Wiley.
21. J. F. Rimbert and J. Pagetti: *Corr. Sci.*, 1980, **20**, 189–210.
22. R. Oltra, G. M. Indrianjafy and R. Roberge: *J. Electrochem. Soc.*, 1993, **140**, 343–347.
23. T. Murata and R. W. Staehle: Proc. 5th Int. Congr. Met. Corros., (ed. N .Sato), 513–518; 1974, Houston, TX, NACE.
24. M. Keddam, R. Oltra, J. C. Colson and A. Desestret: *Corr. Sci.*, 1983, **23**, 441–451.
25. R. Oltra, B. Chapey, F. Huet and L. Renaud: in 'Electrochemical noise measurement for corrosion applications', Montreal, May 1994, (ed. J. R. Kearns, J. R. Scully, P. R. Roberge and D. L. Reichert), ASTM STP 1277; 361–374; 1996, West Conshohocken, PA, ASTM.
26. G. T. Burstein and R. Cinderey: *Corr. Sci.*, 1991, **32**, 1195–1211.
27. G. T. Burstein and A. J. Davenport: *J. Electrochem. Soc.*, 1989, **136**, 936–941.
28. R. Oltra, G. M. Indrianjafy and M. Keddam: in 'Transient techniques in corrosion science and engineering', (ed. W. H. Smyrl, D. D. MacDonald and W. J. Lorenz), Proc.

Electrochemical Society, Vol. 89-1, 363–369; 1989, Pennington, NY, The Electrochemical Society.

29. G. T. Burstein and P. I. Marshall: *Corr. Sci.*, 1983, **23**, 125–137.

30. G. S. Frankel, B. M. Rush, C. V. Jahnes, C. E. Farrell, A. Davenport and H. S. Isaacs: *J. Electrochem. Soc.*, 1991, **138**, (2), 643–644.

31. F. P. Ford, G. T. Burstein and T. P. Hoar: *J. Electrochem. Soc.*, 1980, **127**, 1325–1331.

32. F. Mohammadi, J. Luo, B. Lu and A. Afacan: *Corr. Sci.*, 2010, **52**, 2331–2340.

33. K. Fushimi, K. Takase, K. Azumi and M. Seo: *Electrochim. Acta*, 2006, **51**, 1255–1263.

34. K. Fushimi, T. Shimada, H. Habazaki, H. Konno and M. Seo: *Electrochim. Acta*, 2011, **56**, 1773–1780.

35. T. R. Beck: *Electrochim. Acta*, 1973, **18**, (11), 807–814.

36. J. R. Ambrose: in 'Treatise on materials science and technology', Vol. 23, 175–204; 1983, New York, Academic Press.

37. T. Tsuru, T. Nishimura and S. Haruyama: *Mater. Sci. Forum*, 1986, **8**, 429–438.

38. I. Itagaki, R. Oltra, B. Vuillemin, M. Keddam and I. Takenouti: *J. Electrochem. Soc.*, 1997, **144**, 64–72.

39. G. Rosa, P. Psyllaki, R. Oltra, S. Costil and C. Coddet: *Ultrasonics*, 2001, **39**, 355–365.

40. H. S. Isaacs and Y. Ishikawa: *J. Electrochem. Soc.*, 1985, **132**, 1288.

41. Y. N. Kok, R. Akid and P. E. Hovsepian: *Wear*, 2005, **259**, 1472–1481.

42. D. Deforge, F. Huet, R. P. Nogueira, P. Ponthiaux and F. Wenger: *Corrosion*, 2006, **62**, (6), 514–521.

43. S. Mischler: *Tribol. Int.*, 2008, **41**, 573–583.

44. S. Mischler, A. Spiegel and D. Landolt: *Wear*, 1999, **225–229**, 1078–1087.

45. J. F. Archard: *J. Appl. Phys.*, 1953, **24**, 981–988.

46. E. M. Gutman: 'Mechanochemistry of materials'; 1998, Cambridge International Science, ISBN 9810217811.

47. D. Landolt, S. Mischler and M. Stemp: *Electrochim. Acta*, 2001, **46**, 3913–3929.

48. D. J. Buttle and C. B. Scruby: *Wear*, 1990, **137**, (1), 63–90.

49. R. Oltra, B. Chapey and L. Renaud: *Corr. Sci.*, 1993, **35**, (1–4), 641–646.

50. C. O. A. Olsson and M. Stemp: *Electrochim. Acta*, 2004, **49**, 2145–2154.

Specific testing techniques in tribology: laboratory techniques for evaluating friction, wear, and lubrication

Satish Achanta and Dirk Drees

Falex Tribology NV, Wingepark 23B, Rotselaar 3110, Belgium

sachanta@falexint.com

Friction and wear between contacting surfaces in relative motion is determined by both the intrinsic properties of the materials and by external factors linked to the end users' field conditions. Contacting surfaces between two solid bodies are of interest to two complementary disciplines, namely mechanics and tribology. However, the mechanical approach differs from the tribological one. Indeed, tribologists link measurable macroscopic parameters (such as the coefficient of friction) to geometrical and mechanical characteristics of surfaces (such as roughness, debris) while considering the bulk of the materials as an elastic characteristic. On the other hand, mechanical engineers consider the surface of materials as a transfer zone of load (for example, described as Hertz contact pressure) allowing them to calculate precisely material changes in the vicinity of their surface (elastoplastic behaviour, strain hardening), and the eventual material failure under the prevailing loading conditions (like multi-axial fatigue criteria). In this section, the important factors that influence friction and wear are critically reviewed for both unlubricated and lubricated contacts. Commonly used data analysis methods for the interpretation of friction and wear data obtained from laboratory testing or simulation are presented.

3.1   Friction and wear

Wear can be defined as the surface modification of a component following its movement relative to another body with which it is connected. This phenomenon is detrimental when it begins to affect the aesthetics or functionality of the object that wears. The wear is related to the frictional forces opposing the motion. The origin of these forces has always been the result of research, which is associated with the names of Leonardo da Vinci, William Amontons, Leonhard Euler and Charles-Augustin de Coulomb.

The frictional forces have four main features:

- they vary in proportion to the normal forces between surfaces in contact; the factor of proportionality is called the coefficient of friction μ
- they are independent of the contact area between surfaces
- they are almost independent of sliding speed
- they depend on the materials and their surface state (e.g. roughness, etc.).

Interaction between the forces of friction and wear is essential, since the frictional forces are responsible for energy dissipation in the contact, which generates heat and surface wear. However, there is no systematic relationship between wear and friction and quantification of wear phenomena therefore requires to look in more detail at the mechanisms of wear.

3.1.1   Factors influencing friction and wear

Tribological events such as friction and wear are a system property [1]. Friction and wear result from interactions at the asperity level between surfaces in contact with each other and undergoing a relative motion. Sliding materials are an integral part of the system and the intrinsic properties of these metallic, ceramic, polymeric or composite materials such as hardness, Young's modulus, shear strength, etc., influence friction and wear phenomena. For example, if the difference in hardness between contacting surfaces is large, then the softer material may wear due to two-body abrasion [2]. External mechanical and/or environmental factors that affect friction and wear behaviour, are test parameters such as speed, normal force or stress, temperature, humidity, contaminants, lubrication, etc. Other important factors are wear debris and heat effects generated as a result of mechanical interactions. The nature of the wear debris whether soft/hard, amorphous/crystalline, morphology rolls or angular ones, etc., influences the friction and wear processes, and controls the steady-state conditions [3]. Wear particles have a high surface reactivity, and therefore they further react with their surrounding resulting in reaction products such as oxides, hydroxides, chlorides, fluorides, nitrides, etc., that affect the chemistry of the contacting surfaces. Similarly, thermal effects and heat generated in the contact facilitate reactions, and affect the evolution of friction and wear. As a summary, a comprehensive list of factors affecting friction and wear is given in Figure 3.1.

The reporting of the coefficient of friction and wear data without mentioning the relevant testing or operational conditions is thus meaningless. As a consequence, wear and friction data must be treated and compared with a high degree of caution in such cases. It is known that the reproducibility of friction and wear data obtained by different laboratories may be poor due to differences in concepts and manufacturing of test rigs, and even due to operator's skill or small variations in testing protocols. For example, the relative standard deviation on wear data generated in pin-on-disc tribometers within an inter-laboratory study was between 30% and 40% [4]. Notwithstanding that, high reproducibility in inter-laboratory studies can be achieved by using standardised equipment available in the market. This indicates

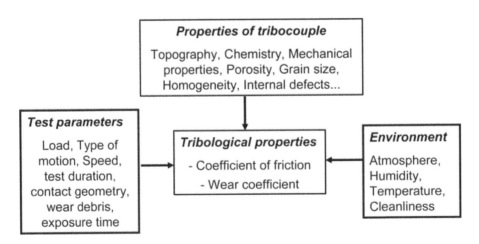

3.1 Commonly known parameters influencing the tribological performance of contacting materials

Table 3.1 List of solid materials commonly used for lubrication

Type of material	Examples	Main lubrication mechanism
Metals	Ag, Pb, Cu, Sn, In [5]	Low shear strength, smearing in the contact
Lamellar solids	Graphite (see Figure 3.2), molybdenum disulphide (MoS_2), tungsten disulphide (WS_2), graphitic boron nitride (h-BN) [6, 7]	Easy sliding of lamellae under shear forces
Oxides	Boron oxide (B_2O_3), CoO, TiO_2 [8]	Low shear strength
Coatings	Diamond-like carbon (DLC), titanium nitride (TiN), tungsten doped DLC (W-DLC), carbon coating (t:a-C)	Transfer mechanism and easily shearing wear debris
Polymers	Polytetrafluoroethylene (PTFE), polyoxy methylene (POM), polyimide, polyacetal [9]	Transfer mechanism

that the repeatability of friction and wear data can be high when well designed and calibrated test equipment is used.

3.1.2   Factors influencing solid and liquid lubrication

Lubrication is mainly done in two ways, namely using solid or liquid lubricants. Solid lubrication can be achieved by coatings, thin films, and other surface modifications such as oxidation affecting the shear strength of the outermost surface of materials. The main purpose of using solid lubricants is to protect contacting surfaces in relative motion in applications where conventional lubricants do not work. Examples are space applications, vacuum related applications, high-temperature applications, etc. An overview of commonly available solid lubricants is given in Table 3.1.

Apart from the materials listed in Table 3.1, lubrication can also be achieved by using hybrid coatings in which a solid lubricant material such as lamellar solids (e.g. Figure 3.2) is introduced as additives or secondary phases into a base material. Examples of hybrid solid coatings are Ni + PTFE, polyimide + graphite, or graphalloy consisting of babbit + carbon fibres.

The environment plays an important role in the efficiency of solid lubricants. For example, MoS_2 oxidises at temperatures above 350°C, and graphite cannot act as a lubricant in a vacuum [11]. Lubrication based on oxides depends strongly on the prevailing experimental parameters because the contact conditions must thermodynamically favour oxidation. This can, for example, be achieved by external heating or frictional heating by applying high speeds or contact loads. Therefore, the choice of a solid lubricant should be made taking into account the operational conditions.

The use of oils and greases is a widely practiced lubrication method in industry. This is because lubrication by oils and greases is more economical compared to solid lubrication. For greases, it is well known that under loading conditions, the oil oozes out of the thickener structure, and lubricates the contact. Therefore, the fundamental lubrication mechanisms by oils and greases are similar. The basic lubrication regimes can be derived from Stribeck curves such as the one shown in Figure 3.3 [12]. Depending on the thickness of the lubricant film between the surfaces, three lubrication regimes can be identified, namely the boundary lubrication regime (regime 1 in

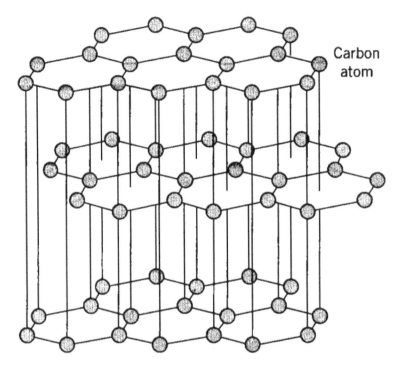

Carbon atom

3.2 Lamellar structure in graphite [10]

Figure 3.3) where asperity contacts between the surfaces occur due to starvation of the lubricant thickness leading to a high coefficient of friction above 0.1 and a high wear in the contacts [13]. This generally occurs at low speed, v, low viscosity, η, and high loads, L. The best lubrication performance is achieved under thin film lubrication conditions (see regime 2 in Figure 3.3) where the surfaces remain separated and the friction is only due to shear in the oil film. Typical values of the coefficient of friction are between 0.03 and 0.08. This regime is termed hydrodynamic lubrication. When the lubricant pressure builds up further, the surface asperities on the contacting materials become elastically deformed, especially when the thickness of the lubricant film is in the order of the surface roughness. This causes the redistribution of pressure in the lubricant film, and this enhances the load bearing capacity of the lubricant. The contacts operating under such conditions are referred to as elastohydrodynamic (EHL) conditions, and the coefficient of friction is lowest below 0.03. At even higher speeds and low loads, the thickness of the lubricant film in the contact becomes large, and viscous drag effects can cause an increase in frictional losses in the contact which is referred to as thick film lubrication (see regime 3 in Figure 3.3). In general, the viscosity of the lubricant is a major factor, and any parameter that affects viscosity will have an influence on friction and wear. For example, test temperature, sliding speed, and loads, affect the viscosity of a lubricant.

3.1.3 Common test configurations

Mechanical components such as linear bearings (bushings, slider bearings, etc.), thrust washers, and specific applications such as piston ring–cylinder bore, operate

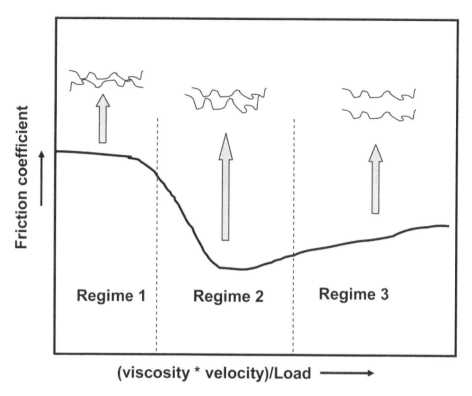

3.3 Different lubrication regimes in a schematic Stribeck curve [12]

under moderate pressure conditions in the range of tens of MPa [14]. Wear in auto-motive engines is in the order of only nm h^{-1} where most lab scale simulations and tests record wear rates in the range of μm h^{-1} up to mm h^{-1} [15]. Often lab scale simu-lations are performed to rank materials for tribological applications by translating real component conditions to lab scale test conditions. This can be achieved by downsizing and simplifying the field conditions. A good simulation requires the choice of the right material couple and test conditions to replicate field conditions in a relevant way. Contact configurations appearing in most industrial parts are summarised in Figure 3.4 [16]. Among these contact configurations, the most com-monly used configuration in laboratory testers is the ball-on-flat contact (Figure 3.5). The main reasons for the widespread use of this configuration are:

(i)   an easy alignment in comparison to a flat-on-flat contact which is a difficult task, especially with small samples
(ii) the results in the case of flat-on-flat contacts can be influenced by edge effects, inclination of the sample, etc.
(iii) the fact that most theories on contact mechanics are developed for ball-on-flat contacts, e.g. Hertzian contact mechanics [17], JKR [18], DMT [19], etc.
(iv) a practical reason, namely the possibility to re-use the same ball by simply turning the ball slightly.

Some disadvantages associated with the ball-on-flat configuration are:

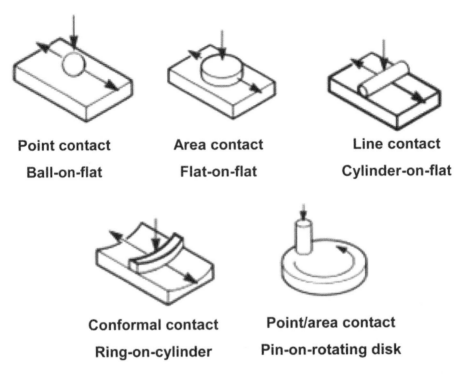

Point contact

Ball-on-flat

Area contact

Flat-on-flat

Line contact

Cylinder-on-flat

Conformal contact

Ring-on-cylinder

Point/area contact

Pin-on-rotating disk

3.4 Most commonly used contacts in tribological investigations [16]

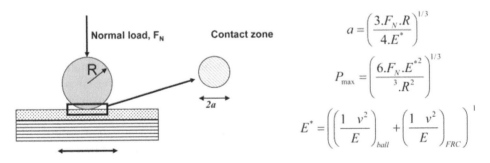

Normal load, F_N Contact zone

$$a = \left(\frac{3.F_N.R}{4.E^*}\right)^{1/3}$$

$$P_{max} = \left(\frac{6.F_N.E^{*2}}{^3.R^2}\right)^{1/3}$$

$$E^* = \left(\left(\frac{1 \; v^2}{E}\right)_{ball} + \left(\frac{1 \; v^2}{E}\right)_{FRC}\right)^1$$

3.5 Left panel: Ball-on-flat test configuration; Right panel Hertzian contact
parameters for elastic contact. *R* is the counterbody radius, *v* is Poisson's ratio,
and *E* is Young's modulus

(i)   the contact pressures can be very high compared to those used under field
       conditions
(ii)   the contact pressure across the contact area is not constant but is at a maximum
       in the axis of the ball and zero at the rim of the contact between ball and flat
       sample
(iii) the relative displacement rate of ball vs. flat sample depends on the diameter of
       the circle of the track that the ball makes on the flat sample
(iv)   the exposure time of the track on the flat sample depends on the angular
       displacement rate of the ball with respect to the flat sample

(v)   the ball surface is continuously in contact with the counter-body during the test

(vi)   finally, wear particles can escape from the contact area relatively easily compared to flat-on-flat contacts. However, one should be aware that a ball-on-flat contact might end up in a different friction and wear behaviour dependent on the position of the flat sample relative to the ball, namely on top of the ball or underneath it, horizontally or vertically positioned!

3.1.4   Interpretation of friction and wear test data

The main risk with the analysis of friction and wear data is that the analysis might be done in a subjective way if one does not take into consideration that the data on friction and wear depend on the measurement technique used. For example, the measurement of the friction force with a sensor requires a non-infinite stiffness that might vary from tester to tester.

One common practice is to record the evolution of *friction force* during the test, and to derive from that the coefficient of friction. This coefficient can be reported as a mean value with its standard deviation as far as one is sure that no main transitions in the contact conditions occur during the test duration. Examples of transitions are running-in phenomena during which asperities present on machined parts are modified, or the formation of wear tracks which drastically modify the contact pressure during tests. The variation in friction data can give ample information on changes occurring in the contact conditions during the tests. For example, in the case of a coated material, the online friction data may reveal the onset of a coating failure by a transition from a low to a high coefficient of friction. This is illustrated in Figure 3.6a, for the case of Al_3Mg_2 coated silicon wafer sliding against corundum under bidirectional sliding conditions.

The *frictional energy loss* can also be determined by integrating the friction force over the distance covered. In the case of a fretting test, a plot of the tangential force versus displacement during each fretting cycle, gives a friction loop as shown in Figure 3.6b. From such a friction loop, information on contact stiffness, running-in stage, wear particles, topography effects (bumps, asperities), material related effects (detection of the presence of different phases in a material), nature of sliding (gross slip, partial slip or stick-slip), etc., can be determined [20].

The analysis of *wear loss* is mostly done after the tribological test. The quantification of wear loss is done in different ways which will be discussed later (Section 3.3.2). The most accessible way to determine wear is by surface observation using microscopic techniques. Light optical microscopy is a first step allowing the detection of changes in morphology or appearance of the worn surface at low magnification. White light interferometry is useful to determine volumetric wear losses. As a second step, scanning electron microscopy is widely used to characterise the wear scar morphology in more detail and at higher magnification, to obtain indications on possible damage mechanisms (fracture, adhesion, abrasion, tribo-oxidation, etc.), and to determine the shape and composition of wear debris. Finally, information on the chemical structure of surface layers, reaction products in the wear track, and composition of the debris can be obtained using XRD, TEM, ToF-SIMS, XPS and other surface analysis tools [21].

In order to minimise the subjective analysis of friction and wear data, one should use *standard test methods* when possible describing the friction and wear testing

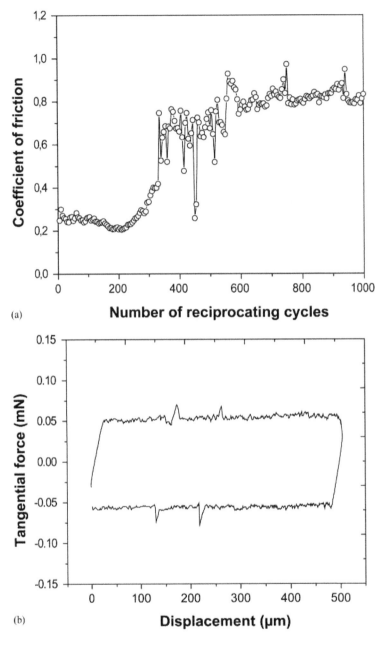

(a)

(b)

3.6 (a) Evolution of friction showing failure of Al$_3$Mg$_2$ coating on silicon wafer after 350 cycles. Test conditions: normal load 100 mN, 1000 cycles, 0.5 mm sliding distance, frequency 1 Hz. Measurement performed in ambient air at 23°C and 50% RH. (b) Friction loop recorded during a reciprocating ball-on-flat test at 500 µN load, 0.5 mm sliding distance, 1 Hz frequency on TiN coated steel sliding against a 5 mm Ø corundum counterbody. Measurement performed in ambient air at 23°C and 50% RH

methodology and data interpretation in terms of internationally agreed standards. For a detailed description of standards, the reader is referred to Chapter 8. Examples are ASTM G133 that relates abrasion testing to reciprocating sliding testing, and ASTM G99 to pin-on-disc testing. The advice of academic and industrial experts in tribology (friction, wear, lubrication) with a background in material engineering and mechanics, complemented with a good insight in surface reactivity of materials, will be of great help in obtaining a correct analysis of tribological test data.

3.1.5   Interpretation of lubrication test data

The testing of lubricants is generally done by standard test methods in which standardised metal specimens are tested in a tribological test configuration such as rotating pin-on-disc, four-ball contact, Pin & Vee block contact or high frequency reciprocating methods. The amount of wear, the maximum load to failure, and the friction force are the most common parameters measured and reported. Standard testing methods generally fall into two categories, namely wear tests and extreme pressure performance tests.

In a **standard wear test**, lubricated contacts are subjected to a fixed set of parameters such as load, speed, temperature and time. The friction force can be measured online, and the wear damage is determined after the test. Well known examples are the ASTM D4172 and ASTM D2266 Four-Ball wear methods to determine the anti-wear properties of lubricating oils and greases. A typical result is shown in Figure 3.7. In this case, lubricant B clearly has a lower friction and lower wear than lubricant A.

To evaluate the **extreme pressure (EP) performance** of lubricants, a lubricated contact is subjected to a progressive or stepwise increase in contact load until seizure occurs. Seizure means that the lubricant can no longer separate the metallic contacts and severe adhesive wear will occur. The load value at which seizure occurs is then used as a measure of the EP property in a particular standardised test. Some standards that aim particularly at the EP properties of lubricants are the Falex Pin & Vee Block method (ASTM D3233) and the Four Ball EP method (ASTM D2783). A typical Pin & Vee Block test result based on ASTM D3233 is shown in Figure 3.8.

3.2   Tribology test equipment

The equipment which enables the measurement of tribological properties such as friction, wear damage and load carrying properties is called a tribometer. Basically, a tribometer will 'rub' (Greek: tribos) two bodies against each other, creating a relative motion. This occurs with a contact load or contact pressure between the bodies and meanwhile the force that resists motion is measured as the friction force. Other relevant parameters that can also be measured are temperatures, acoustic signals, electrical or electrochemical data, etc.

The essential and minimal components of a tribometer are:

– a loading mechanism that applies the contact load between two materials
– a friction measurement system that measures the force resisting relative motion
– a displacement system creating the relative movement between two bodies, and
– holders for the two bodies.

These elements are illustrated in Figure 3.9 for a pin-on-disc tester with description of parts. In this case, one body is typically a rounded pin or ball, and the other body

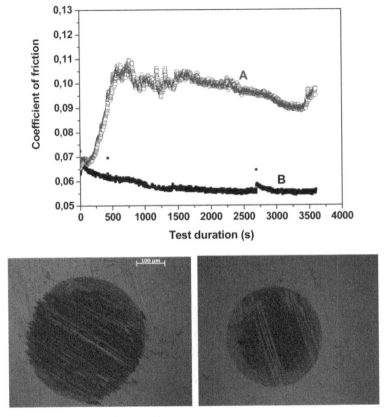

3.7 Difference in anti-wear properties of two different oils A and B tested in a four-ball wear test corresponding to ASTM D-4172: (top) Evolution of the coefficient of friction with test duration; (bottom) wear scar on balls at the end of the four-ball wear tests

a flat plate or disc. This is a popular configuration because of the simplicity of the design and the ease of obtaining materials as balls or flat plates. A ball-on-flat contact also easily self-aligns in one contact point, although this has implications with respect to contact pressure and contact pressure evolution during a test.

A possible classification of state-of-the-art tribometers is presented next. Techniques to investigate friction and wear over a broad range of forces and length scales are discussed, along with a brief insight on the scale dependence of friction and wear.

Note on modularity: Modern commercial tribometers tend to be modular, so that the user could rebuild them from one configuration to another. This is done to reduce the cost of multiple equipment, but the test instruments must be evaluated carefully. Indeed when one instrument must be designed for different configurations, the risk exists that the design is suboptimal for each configuration. One solution to this limitation is the recently introduced tester platform (DS4 tester platform; Tetra GmbH). A large frame has space for up to four different modules that, independently from each other, can create unidirectional ball-on-flat experiments, bidirectional ball-on-flat experiments, torque defined experiments (four-ball, thrust washer), and

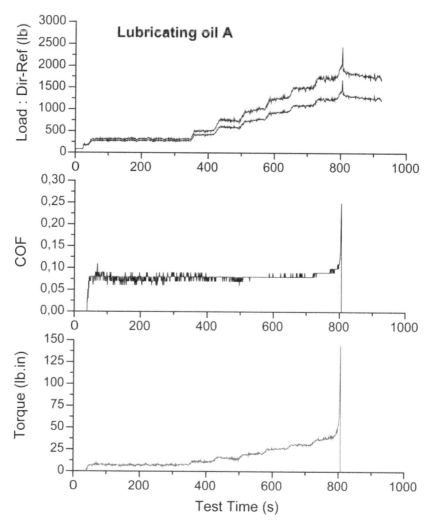

3.8 Typical Pin & Vee block test result of an extreme pressure test on a lubricant. When the load on the contacts reached 2000 lbf (upper graph), a sudden increase in the frictional torque and the coefficient of friction was noticed. This indicates a loss of lubrication at this load, and defines the extreme pressure (EP) load or failure load in this particular standard test

block-on-ring geometry without mechanical compromises. Additionally, characterisation tools, such as a stylus profilometer, a CCD camera, AFM head or scratch tip can be added to the platform. In particular, the efficiency of testing is greatly improved by this approach, because the tested materials can be characterised in-line with the tribological experiment.

3.2.1   Classification of tribometers

The classification of tribological test equipment can be based on many criteria. One is the test geometry that is achieved, and mostly such testers are named after their

1 – Test material

2 – Loading head

3 – Countermaterial

4 – Motion table

5 – Friction force sensor

3.9 Pin-on-disc tribometer (Falex-Tetra DS4 POD)

main test geometry. Examples are the block-on-ring tester, the Falex Pin & Vee block tester, the four-ball tester, the pin-on-disc tester, etc. This classification gives no information on the parameter range during testing. Another way to distinguish tribometers is based on a more elementary view of the tribological contact and takes into account:

–   *the range and scale* of the main test parameters, such as normal forces, sliding distance, speed
–   *the sensitivity of the equipment* – this depends mostly on the resolution of the sensors, and the stiffness and dynamic properties of the whole equipment, and
–   *the mode of displacement* – this is determined by motors or motion systems, and can be unidirectional, linear bidirectional, radial, tangential, etc.

These criteria are visualised and expanded in Figure 3.10.

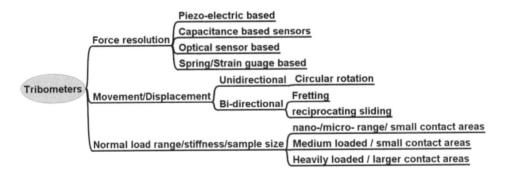

3.10 Detailed classification of the types of tribometers

Range and scale

The normal load range is one of the main design parameters for a tribometer, and it mainly determines the size of test parts and the contact pressures that can be reached in a test. Since the tribometer is a mechanical instrument, the achievable load range often also determines the precision and resolution of load and force measurements.
   Keeping that in mind, the following classification can be made:

– **High load range tribometers** with a normal load range from 100 N to more than 5000 N are usually termed heavy load tribometers. The size of test specimens in such equipment can be large with surface areas over hundreds of mm². The very heavy load tribometers are often unique instruments, purpose-built in research centres and universities.
– **Medium load range tribometers** have a load range of 1 N up to 100 N. These testers use relatively smaller samples, often self-aligned contacts with areas of some hundreds of μm² or less. They are most common in research centres and are used for basic quality control testing in industry because of their limited cost.
– **Low load range tribometers** have a load range from nN up to mN. The contact areas are usually very small, in the range of nm² to tens of μm². Such testers are based on the measurement principles of atomic force microscopes (AFM) and nanoindenters.
– A measurement gap between low and medium load range tribometers is filled by a range of '**meso-load' tribometers** (sometimes mistakenly called nanotribometers) with a load range from some μN up to 10 N. A test machine using a unique deflection-based force sensor is the reference for this load range (Tetra Basalt-MUST, Falex Tribology N.V., Belgium).

The choice of the most appropriate tribometer must take into account various considerations, such as:

– available test sample sizes
– homogeneity of the samples
– fragility of samples
– required contact pressures for simulation
– required measurement precision, and
– required relative displacement.

The two examples hereafter illustrate such a selection process:

1) Self-lubricated composite **bearing materials** consist of a matrix material, filled with lubricating material (e.g. graphite) in a heterogeneous distribution. Such bearings are typically used for large structures where contact pressures are in the 10–100 MPa range. To simulate sliding of this bearing material with a coupon of 100 mm², a load of 10 000 N is required. Often even larger coupons should be tested. This requires the use of a very heavy loaded test machine operating at minimum 1 ton. Conventional strain gauge technology is used as force sensors.
2) For **very thin microelectronic coatings** that can be thinner than 10 nm, a low load, high precision meso-tribometer can be used successfully to characterise the surface. The load is in the micro Newton range, so that friction forces are in the micro Newton range as well, requiring the use of the innovative combination of bending elements and displacement measurement units commonly used in meso-tribometer equipment.

Sensitivity

Tribometers which use piezoelectric sensors have an exceptional force resolution, typically in the range of 10^{-12} to 10^{-10} N. They can be used in lateral force microscopes (LFM) or nanoindenters and are ideal for performing fundamental studies of friction and wear at the nanoscale. They are fragile and have a limited life, and for that reason they are less suited for industrial test equipment. Sensors for meso-tribometers are based on the measurement of displacement of an elastic element, either by optical or laser interferometric sensors or by capacitance measurements. These sensors allow sensitive measurements at a resolution of 10^{-6} N. High load range tribometers use transducers based on conventional strain gauges. Their resolution is typically 0.1% of the full scale measurement range. Recent advances in the use of strain gage technology allow a better resolution of 0.01% of the full scale range.

Displacement mode

Tribometers can also be classified based on their displacement mode. In unidirectional tribometers, the relative motion between two bodies is always in the same direction. Examples are pin-on-rotating disc, block-on-ring, Pin & Vee block or sand abrasion testers. Bidirectional testers are used to simulate fretting mechanisms or reciprocating motion. Many testers can be used both in unidirectional mode and in bidirectional mode, depending on the motor controls and dynamics of the force sensors. In particular, to simulate fretting, a tester must be made as stiff as possible to ensure an accurate relative displacement control, and to limit any resonance effect.

Another classification for displacement mode is to consider the directions of normal force and friction force with respect to the motion axis. In this sense, a block-on-ring configuration is different from a pin-on-disc configuration since the normal load is, respectively, perpendicular and parallel to the axis of rotation.

3.2.2 Measurements over a broad measurement range

The recently introduced meso-tribometers make it possible to study friction and wear uninterrupted over the broadest scales of contact load. This is necessary to understand the origin and scale dependence of friction and wear phenomena, and to explain the large difference between the coefficient of friction measured at the nanoscale (commonly 0.001) and at the engineering scale (usually 0.1 up to 0.8) for a same material combination. In Figure 3.11 [22], the prominent appearance of the meso-tribometer in a previously existing measurement gap is illustrated. This figure focuses on bidirectional experiments but is equally valid for unidirectional experiments.

Bidirectional tribological experiments at the nanoscale were done with an atomic force microscope in the lateral force mode (LFM/FFM). In LFM, a sharp tip typically between 1 and 100 nm radius is brought into contact with a surface, and dragged over a fixed scan size (Figure 3.12a). The deflection or torsion of the sensor is recorded using laser deflection, capacitance, magnetic force, etc. [23]. The working principles of an AFM are detailed in Ref. 24. A lateral force microscopy (LFM) technique for measuring friction forces on graphite and mica surfaces at normal forces of a few nN, was introduced in 1987 [25]. Depending on the stiffness of

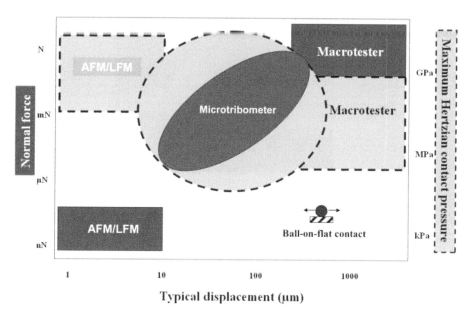

3.11 Force and corresponding pressure range in the case of nano-, micro- and macro-test equipment as collected from literature data

the cantilever, the normal force limits can be chosen. A correct calibration of the equipment is of outmost importance [26] to obtain quantitative results. In most experiments, no or little wear is detected in the contact indicating that LFM measurements are mostly friction tests.

Meso-tribometers (Figure 3.12b) were introduced with operating parameters that fill up a measurement gap between AFM and conventional tribometers. The measurement principle is similar to that of the LFM, namely based on the measurement of the bending of an elastic element. The measuring element is a 25 by 50 mm cantilever with known tangential and normal stiffness. The deflections under load are measured by a set of optical sensors in normal and tangential directions. The force scales can be easily changed by selecting cantilevers with the right stiffness. Unlike in LFM and nanoindenter equipment where the sharp tip creates very high contact pressures, the larger counterbody material that can be fixed to the meso-tribometer sensor results in contact pressures in the MPa range. More information about the equipment principle can be found in Ref. 27.

3.2.3   Scale dependence of friction and wear

The reason to study friction and wear at different load scales is inspired by the large differences found between friction forces in nanoscale and macroscale experiments. The property 'coefficient of friction' is not a material constant but varies strongly with the dominant friction mechanism at a particular measurement scale. It was found that the friction force does not vary linearly with the applied normal load in contradiction to the empirical Amontons' law. Figure 3.13 shows that the coefficient of friction in bidirectional tests between a silicon wafer and a Si_3N_4 sphere is not constant [28]. The variations noticed in the coefficient of friction recorded at different

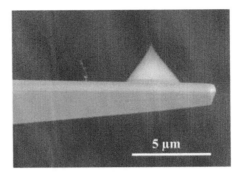

(a)

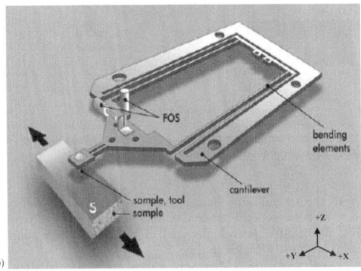

(b)

3.12 (a) Silicon nitride tip 40 nm ∅ used as counterbody in LFM measurements;
(b) Cantilever spring element used in microtribometer (e.g. by Falex Tribology in
its MUST tester, and by CSM in its nanotribometer)

loads can be explained by the relative importance of surface forces that act in a
contact, namely adhesive forces, capillary forces in a humid environment, and
deformation forces.

Figure 3.14 [28] shows different friction mechanisms related to normal load and
roughness. This shows that roughness is important, especially at low normal forces
and low contact sizes, while at higher normal forces and large contact sizes, the
surface roughness is quickly lowered by the destruction of asperities. This creates
wear particles that further influence the friction mechanism.

The scale dependence of the coefficient of friction can be very different for different
materials and environments, but needs to be studied to ensure proper function of
surfaces at various loads or contact stress conditions. The wear mechanisms and wear
coefficients also depend on the measurement scale because the wear coefficient is
not a constant for a given material couple. As the contact size decreases down to the
micrometre scale, the usual wear mechanisms based on plastic deformation, crack
propagation, delamination, fatigue, etc., do not apply anymore. In other words,
the defect-based bulk deformation mechanisms such as dislocation, twinning, etc.,

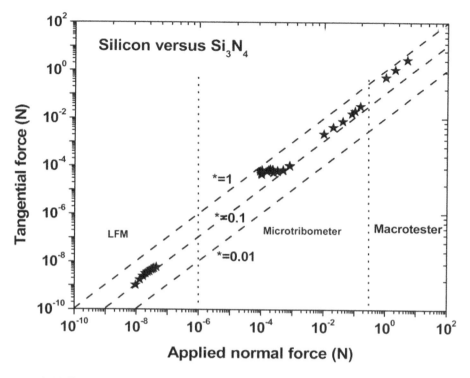

3.13 Tangential force versus applied normal force recorded on silicon sliding against Si_3N_4. Tests done over a broad normal force range in ambient air at 23°C and 50% RH [28]

become less dominant as the contact size decreases. At atomic scales, wear occurs more by transfer of atoms from one surface to the other, a process that is referred to as adhesive wear [29].

Keeping in mind this scale dependence of wear mechanisms, different characterisation methods must be used to achieve a meaningful quantification of wear rate. In the next section, more details on wear mechanisms, and on methods for wear quantification are discussed.

3.3   Wear characterisation methods

Wear is significantly governed by the mechanical properties of a tribocouple. Indeed, degradation mechanisms of materials depend on the intrinsic mechanical properties, and the length scale of deformation. The nature of damage, whether elastic or plastic, in the asperity contact was first explained by Greenwood and Williamson [30]. They introduced the plasticity index which depends on the surface roughness of materials. Later, it was shown that wear modes are significantly affected by the load–roughness dependence of tribocouples [31]. Apart from mechanical parameters, the heat generated in contacts, environment, etc. also affects the wear mode. The different wear modes commonly encountered in mechanical components are frequently discussed hereafter, and a broad overview of wear quantification methods is presented. The quantification of wear can be done using a range of techniques and their choice depends on the extent and type of wear.

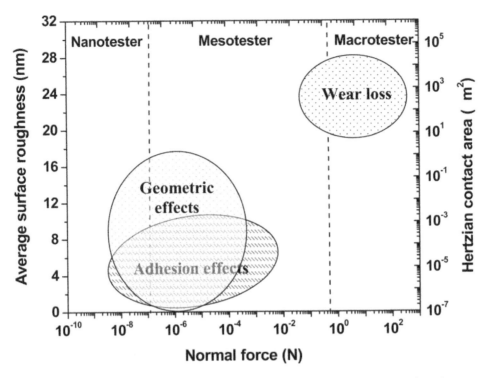

3.14 Schematic representation of various friction mechanisms operating at different ranges of normal force in the case of homogeneous surfaces [28]

3.3.1 Types of wear modes

Depending on the nature of damage, the factors causing wear, and the combination of materials, wear damage on materials can be classified into the following wear modes:

Adhesive wear

This type of wear occurs when asperities on sliding surfaces interact to form cold welds. Adhesive wear is common in metal/metal, metal/polymer, and metal/ceramic contacts. On oxide forming metals, adhesive wear is low as long as the oxide layer remains intact [32]. When the adhesion strength of the weld points exceeds the cohesive strength of the contacting materials, one material is transferred to the counter-face, and adhesion takes place. Adhesive wear in metal/metal contacts is predominantly observed on sliding in vacuum or at high pressure. During continuous sliding, these welds are ruptured. Adhesive wear can be linked to high friction and the presence of lumps of material in the wear track. The best way to confirm adhesive transfer is to use elemental surface analyses. An example is shown in Figure 3.15a for a steel–aluminium tribocouple operating under reciprocating sliding conditions.

Abrasive wear

Abrasion occurs when one of the sliding materials is much harder than its counterpart (Figure 3.15b). The abrasion damage appears like regular patterns of scratches. Typically, abrasive wear is observed when the difference in hardness between the counterfaces exceeds 20% [33]. In such a case, surface asperities on the harder surface can easily abrade the softer material. This type of abrasion is also called two-body abrasion. Abrasion wear is common

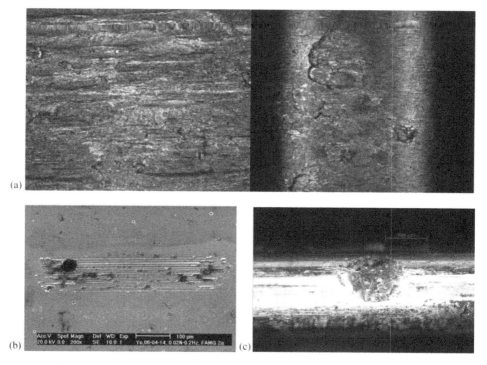

3.15 Examples of wear modes: (a) Adhesive wear. Reciprocating sliding wear test between Al plate against steel cylinder with line-contact (pressure 50 MPa, 10 mm stroke, 1 Hz frequency, and 100 cycles): (left) Al plate, (right) Steel cylinder with Al transfer; (b) Abrasion wear. Reciprocating sliding wear test on Al_3Mg_2 vs. 10 mm Ø alumina ball; at 0.1 N normal force, 1 Hz frequency, 0.5 mm sliding distance for 1000 cycles; (c) Fretting wear. Damage on a gold-coated electrical connector after fretting test at 1 N normal load, 50 μm fretting amplitude, 50 Hz frequency, for 1 million cycles

in material couples such as ceramic/metal and polymer/metal combinations. Depending on the properties of the material, abrasion can take place by severe plastic deformation or by brittle fracture. There are three modes of plastic deformation depending on the attack angle of the hard asperity on softer material, namely cutting, ploughing, and wedge formation. Another kind of abrasion is called three-body abrasion caused by particles confined within the two sliding surface areas. These particles can be wear debris or some hard particles that get trapped in the contact, and start scratching the softer surface. In three-body abrasion, the particle shape and hardness play a major role in the overall wear rate.

Oxidation wear

Oxidation wear is commonly observed on oxide forming materials at high temperatures. In general, oxides on the surface of a metal can be released as wear particles once the oxide layer reaches a critical thickness. Oxidation wear was reported on stainless steel and also on cemented carbides (WC-Co) used on high speed cutting tools [34]. The oxidation wear is much more severe when brittle oxides are formed on metals such as in the case of nickel.

Erosion wear

Erosion wear occurs when hard particles or fluid droplets impinge on a material. The damage occurs as a result of a transfer of momentum. The amount of wear is proportional to particle velocity, impact angle, and density of the material being eroded. Erosion by plastic ploughing and cutting is common in ductile materials. The erosion rate on ductile materials is highest at shallow impact angles (10–30°). Brittle materials such as ceramics erode by microfracturing or excessive cracking, and hence fracture toughness plays an important role in erosion. The maximum erosion rates on these materials often occur under high impact angles (70–90°). The erosion with particles is termed particle erosion whereas erosion caused by the impingement of liquid droplets is called liquid erosion. Liquid erosion is common in steam turbine engines and heavy duty compressors.

Fatigue wear

This type of wear occurs when materials are subjected to cyclic stresses. This damage is common in fretting contacts which vibrate at high frequency over small displacement amplitudes. Under fretting conditions, crack initiation and propagation dominate the wear process. An example of a fretting damage is illustrated in Figure 3.15c. Fretting damage is common in many mechanical components, such as electrical contacts, bearing ring and shaft assemblies, roller bearings, cams, and gears.

Corrosion wear

Corrosion wear is one of the common wear mechanisms observed on components. The sliding action exposes fresh material to its environment and this material further reacts with the corrosive media. The brown debris ($Fe_2O_3 \cdot nH_2O$ or $FeO(OH)$, $Fe(OH)_3$) noticed in tribological tests on low carbon steels is an example of corrosion wear. Dental bracket–wire combinations, the acetabular cup–femoral head in hip implants, undersea drills, and fretting contacts made of reactive materials operating in ambient air, all suffer from corrosive wear. The synergism between wear and corrosion is a complex event known as 'tribocorrosion'. Corrosion wear is one of the most accelerated forms of wear because materials undergo a combined mechanical and chemical attack.

Cavitation wear

Cavitation wear occurs in liquids due to the collapse of gas bubbles at the surface of a material. This results in liquid microjets directed towards the solid surface. The collapsing event can be extremely severe. It may result not only in plastic deformation of the surface due to the high localised pressure, but also in surface fatigue which may cause pitting or erosion. Cavitation wear is common on ship propellers.

3.3.2   Quantification of wear rate

The wear loss can be quantified as a mass loss or as dimensional changes of the tested material. Methods commonly used to quantify wear are where:

- the mass of the material is measured before and after the tribological test, and the wear loss is expressed as a weight difference in µg or mg. This method however, is limited when the wear loss is small and below the resolution of the balance
- the wear volume or wear scar dimensions such as wear depth, width, etc. are measured using stylus-based profilometers or non-contact 3D metrology tools (see Figure 3.16)
- wear is expressed as changes in roughness or asperity level damage (Figure 3.17). The atomic force microscope (AFM) is commonly used for such measurements.

The first definition of wear rate was given by Archard for sliding contacts, and is defined as the volume of material lost per unit normal load and per unit sliding distance. The wear coefficient, K, has units of $mm^3 \, N^{-1} \, m^{-1}$. Typical values of the wear rate for metals are between 10^{-2} and $10^{-5} \, mm^3 \, N^{-1} \, m^{-1}$ [36]. Materials exhibiting a wear rate lower than $10^{-7} \, mm^3 \, N^{-1} \, m^{-1}$ are considered as wear-resistant. Because the

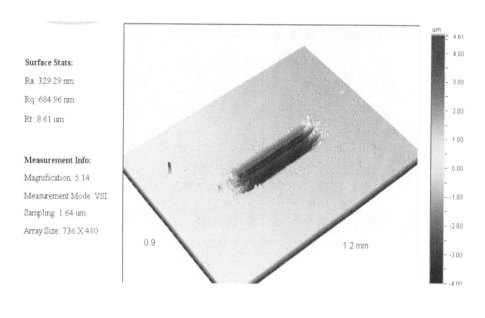

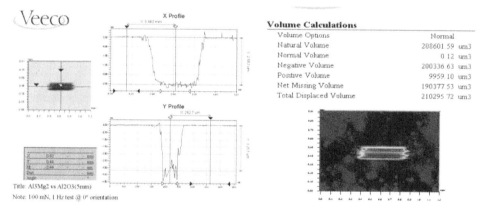

3.16 3D non-contact profilometry of a wear scar obtained after a sliding wear test between Al_3Mg_2 plate and alumina ball at 100 mN, 1 Hz for 1000 cycles

wear mechanisms change with contact size, the wear coefficient is a scale-dependent factor.

Experimentally, the wear rates are calculated by performing a series of tests with different durations, and then for every test, the wear volume is calculated. Later, a scatterplot of wear volume versus corresponding product of normal load and distance is made. The slope of the linear fit to this data is recorded as the wear rate. For example, polyethylene sliding against tool steel during a pin-on-disc test in ambient air has a wear coefficient of 1.3×10^{-7} mm^3 N^{-1} m^{-1} whereas mild steel under the same conditions has a wear rate of 7×10^{-3} mm^3 N^{-1} m^{-1} [36].

Some advanced tribological coatings such as diamond-like carbon exhibit low wear rates in the order of 10^{-10} mm^3 N^{-1} m^{-1}.

The wear rate is also calculated by plotting the wear volume measured at different intervals of test time and by taking the linear slope of such a plot. In some engineering

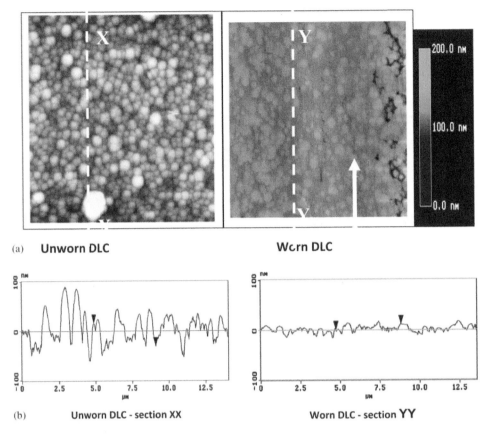

3.17 (a) AFM showing unworn DLC surface (left panel) and topographical changes occurring at the nanoscale [35] on DLC surfaces after 5000 sliding cycles at 100 mN in ambient air at 23°C and 50% RH against Si_3N_4 (right panel; bidirectional arrow indicates sliding direction). (b) Section profiles along unworn DLC section XX (left panel) and worn DLC section YY (right panel). Section locations XX and YY are shown in (a)

applications especially coatings, the wear depth is measured as a function of test duration to measure the coating degradation rate. The wear rate in such cases has dimensions of $mm^3 \ s^{-1}$. In general, volumetric wear is significant during higher friction phases. Therefore wear rate representation as wear volume per unit time can have error if the friction evolves in an unknown manner. This is especially applicable in the case of fretting and bidirectional sliding tests in which speed and acceleration vary during each cycle.

One more popular way to represent wear is by plotting the wear loss versus the corresponding dissipated energy. The dissipated energy from a sliding test is calculated from a friction force, F_t versus displacement plot, x, as $\int F_t.dx$. The slope of the linear fit between wear loss and dissipated energy scatterplot gives the wear rate as shown in Figure 3.18 [37]. It is interesting to note that this has the same dimensions as Archard's wear rate, namely $mm^3 \ N^{-1} \ m^{-1}$. In such an approach, it is assumed that all of the frictional dissipated energy, E_d, is used for the formation of debris.

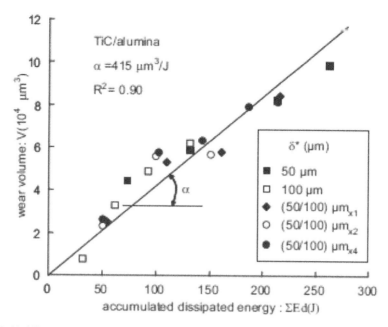

3.18 Wear volume versus dissipated energy for TiC sliding against alumina in ambient air over different sliding distances (load 100 N, frequency 5 Hz). Slope α is the wear rate and corresponds to 415 μm^3 J^{-1} [37]

Table 3.2   Fretting wear rate per unit dissipated frictional energy for different hard coatings sliding against steel or corundum balls under wet or dry conditions [38]

Fretting contact system (flat coated steel disc versus ball)	Wear rate of coating on the flat coated disc based on the dissipated frictional energy, μm^3 J^{-1}
TiN coating *vs.* Cr-steel ball in 0.02 mol L^{-1} Na_3PO_4	30.0×10^3
TiN coating *vs.* Cr-steel ball (dry)	22.0×10^3
TiN coating *vs.* corundum ball (dry)	18.0×10^3
Me-C:H coating *vs.* corundum ball (dry)	5.7×10^3
Me-C:H coating *vs.* steel ball (dry)	4.0×10^3
a-C:H coating *vs.* steel ball (dry)	4.0×10^3
a-C:H coating *vs.* corundum ball (dry)	3.6×10^3
Diamond coating *vs.* diamond coated Si_3N_4 ball (dry)	1.1×10^3
Diamond coating *vs.* corundum ball (dry)	0.4×10^3

The wear rates of some common coatings determined using a wear volume versus energy-based approach for bidirectional sliding conditions are given in Table 3.2. Amorphous carbon coatings appear as promising wear-resistant coatings to resist fretting damage whereas TiN coatings fail [38].

3.3.3   Relationship between friction and wear

The frictional work during sliding can be dissipated in wear particle formation, heating up, and/or a material transformation, etc. [39]. In fact, a general perception is that 'lower friction means lower wear'. Such a relationship is mostly valid for lubricated contacts where a lubricant separates the sliding surfaces. Both friction and wear are then low, e.g. the coefficient of friction is below 0.1 and the wear coefficient is around 10^{-7} mm^3 N^{-1} m^{-1}. However, for unlubricated contacts, such a universal relationship between friction and wear is not noticed. For example, during abrasion, the recorded friction is high because plastic ploughing becomes a major friction mechanism. In the case of a solid lubricant or easily shearing materials, the friction force will be low in the range of 0.04–0.1, but the wear loss can be high, namely in the order of 10^{-3}–10^{-5} mm^3 N^{-1} m^{-1}. Finally, in the case of hard cemented carbides or cermets, the coefficient of friction is moderate, namely 0.2–0.4 while the wear rates remain small. To summarise, there is no fixed relationship between friction and wear processes in unlubricated contacts.

3.4   An industrial tribotesting strategy

In general, research and industrial laboratories that have to evaluate the performance of a mechanical system face a major problem when they want to select an appropriate laboratory test. Hereafter, a strategy is proposed that has already proved its suitability in many industrial cases. For a detailed description of this structured approach to estimate lifetime improvements in components by, for example, design or material selection, readers are referred to Ref. 40.

3.4.1   The Tribological Aspect Number (TAN)

In the 1980s, the IRG-OECD established that tribological properties are a system property [41], and that a material must always be tested or analysed in the system that it works in. That is why any study of a tribological contact must first describe in detail the tribosystem itself.

A **tribosystem** usually consists of two materials in contact with each other undergoing a relative motion, together with an environment in which they operate. This environment can be as simple as a single fluid (lubricant, water) or can be complex and dynamic, namely changing through the lifetime of the components. Apart from the physical environment, there are also environmental parameters to consider, that play a role in the tribological behaviour. Such parameters are, for example, temperature (very important), vibrations, acoustic waves, radiation of any sort, contamination (foreign particles or liquids), etc. In a tribosystem, the materials are interacting with each other under mechanically described parameters such as speed and contact load but materials can also undergo chemical and electrochemical interactions. The most obvious of these last ones is a galvanic contact created in an electrolyte between two metals with different Nernst potentials. An electrochemical reaction (anode/cathode) will then occur, and that can influence the friction and wear behaviour. Indeed mechanical, chemical and electrochemical interactions are surface-related processes that might lead to a surface modification (geometrical, compositional, structural one). Such a surface modification can result in a material degradation or, if well stirred, in a material protection.

As a result, one can already state that just the description of a tribosystem can be very complex. In that respect, it is even clear that the selection of appropriate laboratory testing equipment is complex too. Indeed, it is very difficult to design a good laboratory test that takes into account all of the system properties. Usually, it is even impossible or economically impractical. One should also keep in mind that laboratory tests are, by their nature, simplifications of the real components, and that they aim at an accelerated testing of the functionality. However, they offer the advantage that tests can be done under better controlled conditions than on components in real assemblies. Results can often be obtained faster, and more properties can be measured in laboratory tests. These advantages outweigh the disadvantage that a laboratory test can never be done exactly under the same operational conditions as a real component in the field.

When selecting a proper laboratory test, a number of considerations have to be made, and they can be in the form of questions:

1. Can we design an appropriate mechanical simulation?
2. What are the required results (tribometrics)?
3. Can we consider correctly the contact pressures?
4. Can we allow accelerated testing?
5. Is it possible to perform simultaneous testing?
6. Can we obtain an insight into wear evolution rather than using absolute wear rate?

Among these considerations, the design of a correct mechanical simulation is not that straightforward. In that respect, the '*Tribological Aspect Number*' or TAN approach, provides us with an easy to apply tool to obtain a mechanical description of a field system expressed as a 4-digit number (see Section 3.4.2).

3.4.2 Background and questions

Can we design an appropriate mechanical simulation?

The 'Tribological Aspect Number' approach is an attractive way to answer this question by considering four main parameters for the two contacting parts in an engineering system. These parameters are speed, contact area, contact pressure, and lubricant entry angle. To initiate the design of an experiment, the typical evolution of each of these parameters must be determined, before assigning numerical values to them.

Speed

This can be distinguished in unidirectional relative motion, oscillating (or reciprocating) motion, roll-slip combination, and fretting (small displacement oscillation) as shown schematically in Figure 3.19 [42].

Contact area evolutions

These are considered, ranging from a stable point contact (e.g. in rolling ball bearings) to a constant area contact (flat-on-flat, conforming surfaces), and to an open or closed spiral contact. Each of these evolutions (Figure 3.20) have different effects on the contact in terms of the exposure time of materials to the environment in between successive contact events, wear particle behaviour, evolution of contact pressures, etc. In the case of tribocorrosion, the effect on exposure time is certainly one of the most important.

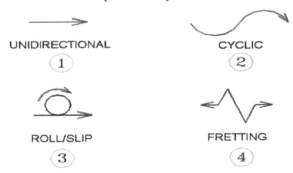

3.19 Schematic overview of the main parameter 'speed' in a tribological system [42]

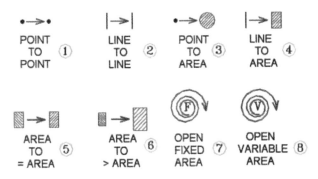

3.20 Schematic overview of the main parameter 'contact area evolution' in a tribological system [42]

Contact pressure

This is determined both by the evolution of the contact area, and by the external loads applied on the contact. In this category, it is important to distinguish between unidirectional evolving contact pressure (increasing or decreasing), cyclic loading–unloading mechanisms or (high frequency) loading around an average pressure (Figure 3.21). The cyclic pressure variations can cause fatigue, not present in a unidirectional evolving contact pressure. The difference between full loading–unloading and cyclic oscillating pressure around an average, can be very relevant especially when the environment can influence the material surface. For instance, full unloading in an electrochemical environment will expose the complete contact area to the environment while under a cyclic oscillating pressure. There is always a part of the contact area that remains unexposed to the environment.

Entry angle

The final category is the '*entry angle*' (Figure 3.22) which is most relevant for applications in a fluid. The entry angle determines the lubrication regime together with speed

CONTACT PRESSURE/LOAD CHARACTERISTIC

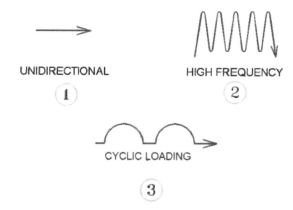

3.21 Schematic overview of the main parameter 'contact pressure' in a tribological system [42]

and load. Components with a small entry angle have a better chance of reaching (elasto) hydrodynamic lubrication and thus leaving the boundary lubrication regime. As the materials become separated by a fluid film, the whole contact process is evolving.

When a field application has been satisfactorily defined with the above four categories, we obtain a 4-digit number (TAN number) describing the nature of the contact. A few examples of the TAN-number for moving parts in a complex engineering system are shown in Figure 3.23.

Once a good TAN-description of a given field problem has been made, it becomes easy to design or to choose a proper lab test geometry by selecting a lab test set-up with the same TAN-code ensuring a conforming contact simulation.

The lab testing strategy to be followed when one wants to apply the TAN-approach, is summarised in Figure 3.24.

ENTRY ANGLE CHARACTERISTIC

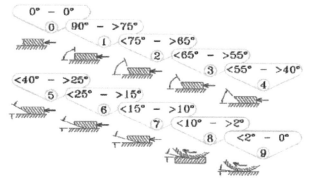

3.22 Schematic overview of the main parameter 'entry angle' in a tribological system [42]

TYPICAL TAN FIELD CODES

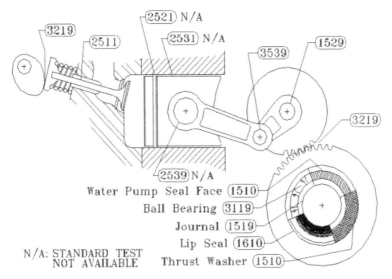

N/A: STANDARD TEST
NOT AVAILABLE

Water Pump Seal Face (1510)
Ball Bearing (3119)
Journal (1519)
Lip Seal (1610)
Thrust Washer (1510)

3.23 TAN-numbers for the constituting moving parts in a complex engineering system [42]

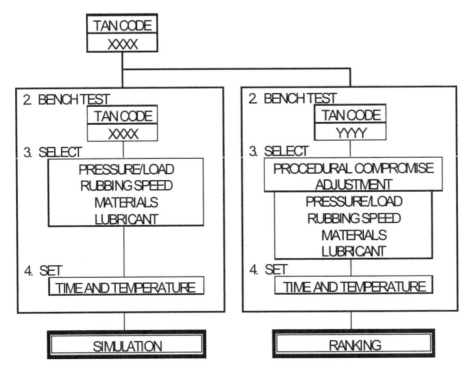

1. IDENTIFY FIELD PROBLEM

TAN CODE
XXXX

2. BENCH TEST
TAN CODE
XXXX

3. SELECT
PRESSURE/LOAD
RUBBING SPEED
MATERIALS
LUBRICANT

4. SET
TIME AND TEMPERATURE

SIMULATION

2. BENCH TEST
TAN CODE
YYYY

3. SELECT
PROCEDURAL COMPROMISE
ADJUSTMENT
PRESSURE/LOAD
RUBBING SPEED
MATERIALS
LUBRICANT

4. SET
TIME AND TEMPERATURE

RANKING

3.24 Flow chart for a TAN-approach [40]

The next step to reach a good simulation or ranking of materials is the correct selection of the parameters and tribometrics. For this part, the questions mentioned above should be answered as completely as possible. It is at this stage that experience with testing and with the industrial component is important.

A few notes on answering the above questions are given next as examples, but for every application, they need to be repeated and answered in as much detail as possible.

What are the required results: tribometrics?

Depending on the application, the required test results may differ. For instance, if an application is studied where the friction needs to be lower, then it is obvious that the friction force must be measured in a reliable and precise way. On the other hand, the functionality of an electrical connector is based on its electrical conductivity, so in a friction test for electrical connectors, the electrical resistance is a needed tribometric.

Can we consider correctly the contact pressures?

This is perhaps one of the most important aspects to achieve a reliable tribotesting. The test should be performed at the same contact pressure range as the application to ensure conformity and to reach a good correlation. It makes no sense to increase the contact pressure to accelerate the testing when the increased pressure becomes unrealistic and induces a friction and wear mechanism different from the one in the real application. Sometimes, the use of an increased pressure can be used as a quality control test, for instance for lubricants, but it is rarely useful to do this for a real application simulations.

As an example, we can consider the widespread 'pin-on-disc' method for testing the wear resistance of hard coatings. Producers of and researchers on such coatings often use a small ball on a flat test sample operated at a normal load of only a few Newtons. But on hard materials, even a 1 N load can generate Hertzian contact pressures of more than 1 GPa or even 2 GPa. Very few applications, with the exception of ball bearings or certain gears, are operated at such high contact pressures. The following question arises: why tests are nevertheless often done at such high loads? The answer to this brings us to the next question.

Can we allow an accelerated testing?

Accelerated testing is of course desirable to obtain test results and to derive conclusions in a timescale shorter than the lifetime of components in the field. A typical machine part (engines, machines) should have a lifetime of 1000s of hours. It is not desirable to repeat that in laboratory tests. So, an acceleration should be attempted, e.g. by increasing test temperature, normal load or speed, or by focusing only on the most damaging wear mechanism in a component. In that case, it must be remembered that the accelerated test may not produce a completely different wear mechanism than in the field application.

The following example illustrates this. When a machine component has an acceptable wear (loss of tolerance, for example) of 10 μm, and the lifetime must be 5000 h, then a wear rate of 2 nm per hour is acceptable. The testing challenge is to perform a wear test that reproduces this wear rate, and at the same time is able to measure such low wear after a short test time. A 10 h test would only produce 20 nm, which is extremely difficult to measure confidently. Accelerated testing can only be allowed to the extent that no change of wear mechanism occurs. This implies that in the test programme, first the most optimal parameters have to be found that combine realistic wear mechanism with some acceleration. A 'tuning' of test parameters is often needed before routine tests can be done.

Is it possible to perform simultaneous testing?

A major drawback of many lab test machines is that they usually consist of a single station friction or wear tester without incorporated characterisation tools. This makes lab testing very expensive: a single station test machine can be occupied for many days, even weeks, to produce one single test result. The test sample then has to be physically transported from the test machine to a characterisation tool (e.g. microscope, profilometer, balance) and cannot be replaced in the tester for a continuing test.

This lack of efficiency works against the discipline of lab testing, especially with the emphasis on reliability analysis of test results. The obvious solution is to use equipment that allows simultaneous testing (multi-station) or equipment that incorporates automated characterisation tools. 'Combinatorial testing' as it is known in the pharmaceutical and chemical industries, would be a major step forward in the efficiency and reliability of tribological lab testing. Fortunately, some multi-station and combinatorial equipment are becoming available in the market (e.g. Tetra GmbH, Ilmenau (Germany) DS4 equipment platform with integrated characterisation tools; Phoenix Tribology (UK) 50 and 100 station wear testers, originally for biomaterials testing).

Can we obtain an insight into wear evolution rather than using absolute wear rate?

A lab test remains an approximation of reality, so it is best to consider relative wear rates rather than absolute wear numbers as an interpretation of material behaviour. *A priori*, absolute wear in a tribometer has no correlation with an application, especially if the tribometer test is accelerated to some extent. A relative comparison of wear rates for different materials or lubricants gives useful information. Equally important is to obtain information about the wear evolution of a material. The 'running in' of materials can be a deciding factor for the total lifetime, and the typical evolution contains more information than a single wear volume result at the end of a single test. This is illustrated in Figure 3.25 in a simplified form. A material that follows a wear evolution according to the broken line will have a faster running-in, but in the long run, it will wear less than the material corresponding to the full line. Depending on the test time (shorter or longer than the running-in time), a correct or misleading conclusion could be drawn.

3.5   Conclusions

Friction and wear mechanisms are determined by both intrinsic properties and the external (experimental or operating) conditions of the tribosystem. In particular, the

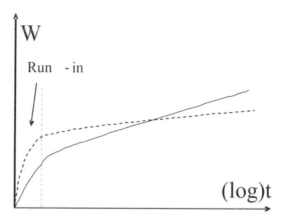

3.25 Schematic diagram showing evolution of wear loss, *W*, versus test duration, *t*

environment in which the sliding occurs has a significant influence on both friction and wear. Under unlubricated conditions, there is no fixed relationship between friction and wear. Both the coefficient of friction and the wear rate are dependent on the measurement scale, and they are not constants for a given material couple.

For a good simulation and characterisation of friction and wear, experimental parameters representing the contact conditions during field operation should be chosen carefully. Many studies have shown that, depending on the chosen simulation parameters, friction and wear modes can evolve causing misleading interpretations.

The existing state-of-the-art tribometers can cover a wide range of force, length and time scales. The choice of tribological test equipment depends on the measurement sensitivity required for friction, and the extent of wear damage on the surfaces. Because most tribometers suffer from poor repeatability/reproducibility issues, standard test equipment and test methods (ASTM, ISO, DIN) should be used when setting up a material/lubricant ranking process. The standard test methods allow us to compare data from different laboratories within defined error limits. In this way, the interpretation of the data can be made with higher confidence.

The determination of this wear evolution is again a time consuming process. It involves running a series of wear tests with increasing duration and characterisation of the wear loss after each test. Alternatively, using a method that can measure 'on-line' wear would be a great improvement. Thin Layer Activation (TLA) can be used in lubricated contacts: radioactive tracer elements are present in the material by activating the surface. When the material wears, the particles can be detected in the circulated oil flow. In a simple 'pin-on-disc' experiment, it would be possible to perform intermittent wear track measurements (e.g. profilometry) in an 'in-line' way. This means performing them on the test machine without removing the disc so that the experiment can be continued. This method also improves the efficiency of lab testing enormously. Some emerging test platforms have the capability of both multi-station testing and in-line characterisation.

References

1. D. Dowson: 'History of tribology'; 1979, London, Longman.
2. D. H. Buckley: Tribology Series, No. 5; 1981, Amsterdam, Elsevier.

3. I. L. Singer, S. D. Dvorak, K. J. Wahl and T. W. Schraf: *J. Vac. Sci. Technol. A*, 2003, **21**, S232–S240.
4. 'Standard test method for wear testing with a pin-on-disk apparatus', ASTM G99-05; 2010, West Conshohocken, PA, ASTM International.
5. I. L. Singer: in 'Fundamentals of friction: macroscopic and microscopic processes', (ed. I. L. Singer and H. M. Pollock), NATO Series E: Applied Sciences, 1992, **220**, 237–261, ISBN 0-7923-1912-5.
6. X. Zhang and J.-P. Celis: *Appl. Surf. Sci.*, 2003, **206**, 110–118.
7. M. Kawaguchi, K. Shinya and M. Yasuji: *J. Phys. Chem. Solids*, 2008, **69**, 1171.
8. E. De Wit, L. Froyen and J.-P. Celis: *Wear*, 1999, 116–123.
9. J. K. Lancaster: *Tribology*, 1972, **5**, (6), 249–255.
10. 'Structure of lamellar solids', http://www.benbest.com/cryonics/lessons.html.
11. L. Rapoport *et al.*: *Nature*, 1997, **387**, 791–793.
12. 'ASM Handbook, Friction, wear, and lubrication technology', Vol. 18, (ed. P. Blau); 1992, Materials Park, OH, ASM International. ISBN 0-087170-380-7.
13. M. M. Maru, D. K. Tanaka and J. Braz: *Soc. Mech. Sci. Eng.*, 2007, **29**, 55–61.
14. 'The tribology handbook', (ed. M. N. J. Neale), 2nd edition; 1997, Oxford: Butterworth-Heinemann, UK. ISBN 0750611987.
15. M. Scherge, D. Chakhvorostov, and K. Pohlmann: *Tribologie Schmierungstechnik.*, 2003, **50**, 5–9.
16. Courtesy of Phoenix Tribology, www.phoenix-tribology.com.
17. H. Hertz: *J. Reine Angew. Math.*, 1881, **92**, 156.
18. K. L. Johnson, K. Kendall and A. D. Roberts: *Proc. R. Soc. Lond. A*, 1971, **324**, 301.
19. B. V. Derjaguin, V. M. Muller and Y. P. Toporov: *J. Colloid Interface Sci.*, 1975, **53**, 314.
20. H. Mohrbacher, J.-P. Celis and J. R. Roos: *Tribol. Int.*, 1995, **28**, 269–278.
21. K. Miyoshi: 'Surface characterization techniques: an overview', NASA/TM 2002-211497; 2002, NASA.
22. S. Achanta: 'Investigation of friction from nano to macro force scale under reciprocating sliding conditions', PhD thesis, K.U. Leuven, Leuven, Belgium, April 2008. ISBN: 978-90-5682-914-8.
23. M. L. B. Palacio and B. Bhushan: *Crit. Rev. Solid State Mater. Sci.*, 2010, **35**, 73–104.
24. E. Meyer, H. J. Hug and R. Bennewitz: 'Scanning probe microscopy'; 2004, Springer, ISBN: 978-3-540-43180-0.
25. C. M. Mate, G. M. McClelland, R. Erlandsson and S. Chiang: *Phys. Rev. Lett.*, 1987, **59**, 1942.
26. E. Liu, B. Blanpain and J.-P. Celis: *Wear*, 1996, **192** (1–2), 141–150.
27. S. Achanta, T. Liskiewicz, D. Drees and J.-P. Celis: *Tribol. Int.*, 2009, **42**, 1792–1799.
28. S. Achanta and J.-P. Celis: *Wear*, 2010, **269**, 435–442.
29. R. Colasço: in 'Fundamentals of friction and wear at the nanoscale', (ed. E. Meyer and E. Gnecco); 2007, Springer, ISBN: 978-3-540-36806-9.
30. J. A. Greenwood and J. B. P. Williamson: *Proc. R. Soc.*, 1966, **A295**, 300–319.
31. E. Vancoille, J.-P. Celis, J. R. Roos and L. M. Stals: Proc. 18th Leeds-Lyon Symposium on 'Wear particles: from the cradle to the grave', Lyon, France, 1991, Amsterdam, Elsevier.
32. D. Tabor and F. Bowden: 'Friction and lubrication of solids'; 1968, Oxford, Clarendon Press.
33. A. van Beek: 'Advanced engineering design'; 2009, TU Delft, ISBN-10: 90-810406-1-8.
34. H. S. Shan and P. C. Pandey: *Wear*, 1976, **37**, 69–75.
35. S. Achanta, D. Drees, and J.-P. Celis: *Wear*, 2005, **259**, 719–729.
36. I. M. Hutchings: 'Tribology: Friction and wear of engineering materials'; 1992, London, Edward Arnold.
37. S. Fouvry, C. Paulin and T. Liskiewicz: *Tribol. Int.*, 2007, **40**, 1428–1440.

38. J.-P. Celis, L. Stals, E. Vancoille and H. Mohrbacher: *Surf. Eng.*, 1998, **14**, (3), 205–210.
39. D. Shakhvorostov *et al.*: *Wear*, 2007, **263**, 1259–1265.
40. M. Anderson: in 'Bench testing of industrial fluid lubrication and wear properties used in machinery applications', (ed. G. E. Totten, L. D. Wedeven, M. Anderson and J. R. Dicke), STP 1404, 283–295; 2001, West Conshohocken, PA, ASTM International.
41. H. Czichos and K. H. Habig: 'Tribologie-Handbuch'; 2010, Wiesbaden, Germany, Vieweg+Teubner-Verlag, ISBN 978-3-8348-0017-6.
42. D. Drees and J. P. Celis: Proc. 13th Int. Colloquium Tribology, 'Lubricants, materials and lubricant engineering', 15–17 January 2002, Stuttgart/Ostfildern, Germany, (ed. W. J. Bartz), Vol. 1, 409–411; 2003, Esslingen, Germany, Technische Akademie.

4

Specific testing techniques in tribology and corrosion: *Electrochemical techniques for studying tribocorrosion processes in situ*

Vincent VIGNAL

Université de Bourgogne, I.C.S, UMR CNRS 5613, F-21078 Dijon Cedex (France)
Vincent.Vignal@u-bourgogne.fr

François WENGER

Ecole Centrale Paris, Dept. LGPM, F-92295 Châtenay-Malabry (France)
Francois.wenger@ecp.fr

Bernard NORMAND

INSA- Lyon MATESIS, UMR CNRS 5510 R12S, F-69621 Villeurbanne Cedex (France)
Bernard.normand@insa-lyon.fr

The mechanism of tribocorrosion is not yet fully understood, partly due to the complexity of the chemical, electrochemical, physical, and mechanical processes involved. In addition, synergistic effects increase the complexity of the system. Mechanical and electrochemical in situ measurements are needed to obtain information about the synergistic and antagonistic mechanisms. They must be done under strictly well controlled conditions using materials that were previously well characterized (microstructure, surface roughness and chemical composition, etc). Various electrochemical methods like open-circuit potential measurements, polarization curves, and EIS measurements can provide essential information on the tribocorrosion mechanisms, as well as on the kinetics of reactions, the existence of a protective passive film, and the interactions between electrochemical reactions and friction. In this Chapter, an overview of the techniques suitable for a mechanical-electrochemical approach at the macroscale and microscale are presented. The capabilities and present limitations of the electrochemical techniques for studying tribocorrosion processes are addressed. The challenges in multi-scale and multi-disciplinary approaches are also discussed in this section.

4.1 Introduction

Over the past years, an increasing interest is noticed in literature on the possibility to investigate the combined corrosion–wear degradation of materials by electrochemical methods. The influence of passivity on the tribocorrosion of carbon steel and TiN coatings in aqueous solutions was investigated [1–2]. Potentiodynamic polarization measurements and tests performed under electrochemical control at different potentials were used. The influence of potential on the tribocorrosion of nickel and iron in sulfuric acid solutions was reported [3]. Watson *et al.* [4] reported on methods for measuring corrosion–wear synergism. A microelectrochemical technique in which the current of a metal probe in an electrolyte under rubbing conditions is measured, was proposed [5] to study the corrosion–wear synergy. Finally, it is worth mentioning that some modelling of tribocorrosion has appeared in literature recently. Jemmely *et al.* [6] proposed an electrochemical modelling of passivation phenomena in tribocorrosion, while Garcia *et al.* [7] analyzed the corrosion–wear of passivating materials in sliding contacts based on the concept of an active sliding track area. The immersion of a material in an electrolyte allows an *in situ* characterization of the surface state of that material and its evolution during sliding tests. That surface state and reactivity can be determined by different electrochemical techniques such as open circuit potential measurements, polarization measurements and impedance measurements. The information obtained by such *in situ* electrochemical techniques can be linked to *in situ* mechanical measurements, such as friction force recorded during sliding tests.

The concept of the tribometer must be compatible with the installation of the test sample in an electrochemical cell containing the electrolyte and the electrodes which are required for performing electrochemical measurements under suitable conditions. The "pin-on-plane" test set up appears well suited for evaluating the resistance to tribocorrosion of passivating metals or alloys, and to highlight the role of the passive film. Indeed, if the surface of the sample is the surface of a passivating material to be tested, the pin partially or totally destroys the passive film in the contact area. After the passage of the pin, the passive film tends to re-grow before the subsequent contact event. At each passage of the pin, the material undergoes a mechanical action applied on a partially depassivated surface and also an electrochemical action leading to the dissolution and repassivation of the bare metal.

In contrast to reciprocating (bidirectional on the disk material) contact conditions, unidirectional sliding conditions lead to an electrochemical steady-state by averaging [8–9]. Such a dynamic steady state is however quite atypical since between two successive passes of the pin/ball, a point at a given fixed position on the sliding track reacts with the surrounding liquid, and is therefore time-dependent. Notwithstanding that evolution of the material in the sliding track with time, the achievement of such a dynamic steady state condition is required for the implementation of some electrochemical techniques like polarization curves, impedance measurements, etc.

A wide variety of electrochemical techniques are available to study the corrosion and passivation involved in tribocorrosion. These techniques can be used not only to identify the nature of the electrochemical reactions involved in tribocorrosion and their mechanistic role, but also to quantify the contribution of these reactions to the total material loss. Among the techniques most commonly used, it may be mentioned:

- Measurement of the open circuit potential (OCP) and of its variations with time.
- The plot of polarization curves.
- The plot of electrochemical impedance diagrams (EIS).
- The so-called "potential step" method [10–11].
- The electrochemical noise analysis.

Hereafter, these different electrochemical techniques are illustrated for two types of passivating materials , namely a stainless steel AISI 316 L exhibiting a very quick passivation in the test solution used, and an alloy Fe–31% Ni having a very low passivation rate in the test solution used.

4.2   Open circuit potential measurements (OCP)

4.2.1   OCP evolution

Classically, OCP follows several variations with immersion time as the ones shown schematically in Figure 4.1. These time evolutions reflect the balance between anodic and cathodic reactions. The OCP is a thus a mixed potential. For instance, in the case of a passivation process there is an increase in the potential in the so-called nobler direction (as in curve *a*). Curve *b* is characteristic of an activation of the material that undergoes a uniform corrosion. Curves *c* and *d* characterized by a potential drop on immersion are representative for cases where a surface evolution is required to achieve a film growth (curve *c*) or when a depassivation phenomena appear just after immersion(curve *d*).

Under sliding, a galvanic coupling between the material in the sliding track area (worn part) and the material outside the sliding track area (unworn part) on the disk surface may take place [9]. Consequently, the open circuit potential will be determined by the following parameters:

- The respective intrinsic open circuit potentials of the materials in worn and unworn areas. These open circuit potentials are different because the electrochemical state of the metal is disturbed by wiping of the surface films that may consist of adsorbed species, passive films, or corrosion products, in the sliding contact, and by a mechanical straining of the metal.
- The ratio of worn to unworn areas. In particular, if the extent of the worn area increases, the open circuit potential of the disk will shift depending on the controlling electrochemical processes, being either the anodic (e.g., the dissolution of the metal) or the cathodic reaction (e.g., the reduction of hydrogen ions or protons and/or dissolved oxygen).

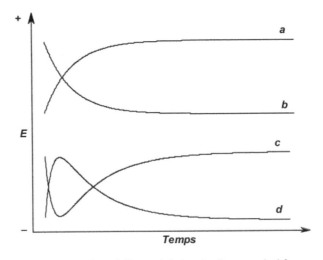

4.1 Evolution of Open Circuit Potential classically recorded for passive materials (*a, c*) and active material (*b, d*)

- The relative position of worn and unworn areas. As a result of the galvanic coupling, a current is flowing between anodic and cathodic areas. The ohmic drop in the test solution may induce a non-uniform distribution of potential and current density over the disk surface. The measured open circuit potential is thus an average value depending on that distribution.
- The mechanism and kinetics of the anodic and cathodic reactions in worn and unworn areas.

4.2.2   Application to AISI 316 stainless steel

The variation of the open circuit potential with time before, during, and after a sliding test is shown in Figure 4.2 for stainless steel AISI 316 (flat disk finished with emery paper grade 200) subjected to uni-directional sliding against a corundum ball (diameter 10 mm) both immersed in 0.5M H_2SO_4.

Before starting the corrosion–wear sliding tests, the freshly ground and thus active AISI 316 stainless steel disk material was immersed for 10^4 s. In that time interval, a large increase of the open circuit potential of the stainless steel is noticed. This increase indicates that a stable passive surface state is achieved. At the time the corundum ball is loaded on the rotating stainless steel disk, a sudden decrease of the open circuit potential takes place (see cathodic shift at 10^4 s). The open circuit potential during the corrosion–wear test is quite close to the open circuit potential of the active base material noticed on immersion of the sample in the electrolyte. Finally on unloading of the ball from the disk, the open circuit potential of the stainless steel disk starts to increase (anodic shift) and reaches after some time the initial open circuit potential back. This indicates the re-establishment of a kind of passive state on the surface of the stainless steel material in the sliding track area. It should be noticed that the rate at which the open circuit potential decreases on loading and increases on unloading is totally different. This indicates that the underlying processes are

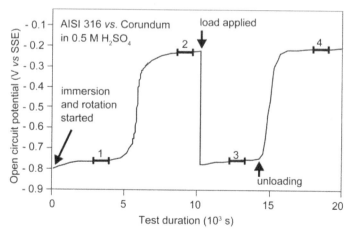

4.2 Variation of the open circuit potential of a stainless steel disk immersed in 0.5 M H_2SO_4 before (e.g. time intervals 1 and 2), during (e.g. time interval 3), and after loading (e.g. time interval 4) against a corundum ball. The disk was continuously rotated at 400 rpm and the load was 10 N. Sliding speed was 0.63 m s^{-1} [28]

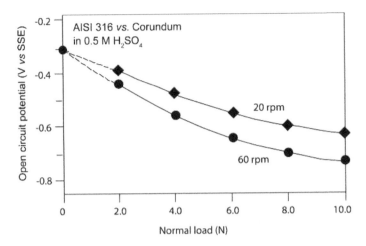

4.3 Variation of the open circuit potential with sliding parameters (normal load, contact frequency) for AISI 316 stainless steel immersed in 0.5 M H_2SO_4. Counter-body: corundum pin. Sliding speed at 20 rpm was 0.031 ms^{-1}, and at 60 rpm 0.093 ms^{-1} [28]

different in both cases and have different kinetics. Indeed on loading, a sudden mechanical destruction of the passive surface film is taking place, while on unloading repassivation takes place at a rather limited oxidation rate.

The open circuit potential of passive materials in sliding contacts is quite sensitive to the loading conditions. So, for example, the open circuit potential of AISI 316 stainless steel disks immersed in sulfuric acid, varies largely with normal force and contact frequency (see Figure 4.3).

The shift in the open circuit potential on loading can be explained as follows:

- The passive film can be partly destroyed (cracking, partial removal) in the contact area under sliding conditions where friction occurs. This may initiate a galvanic coupling between the passive surface layer and the bare base material, with a local dissolution of the base material as a consequence. Since the open circuit potential of active material is cathodic compared to passive material, the open circuit potential of partially depassivated materials, shifts towards the active region.
- The contact area (e.g. the Hertzian contact area under conditions of elastic loading) between first bodies increases at increasing normal load. This generates a larger sliding track area. As a consequence, a larger area of active material in the sliding track causes a lowering of the open circuit potential.
- The variation of the open circuit potential with contact frequency is linked to the time interval in between two successive contact events during which material in the sliding track may repassivate. At higher contact frequency, the amount of active material in the sliding track area thus increases. This results in a cathodic shift of the open circuit potential due to an increased ratio of active-to-passive area.

4.2.3   Application to Fe–31% Nil

In analogy, the variation of the open circuit potential before, during, and after sliding test is shown in Figure 4.4 for Fe–31% Ni (flat disk finished with emery paper grade

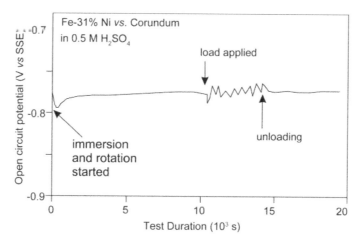

4.4 Variation of the open circuit potential of a Fe–31% Ni disk immersed in 0.5 M H_2SO_4 before, during loading at 10 N, and after unloading. A flat-ended corundum pin was continuously sliding at 80 rpm. Sliding speed was 0.063 ms^{-1} [28]

200) subjected to a uni-directional sliding against corundum balls (diameter 10 mm) in 0.5M H_2SO_4.

The open circuit potential is not altered significantly by the establishment of a sliding contact. A galvanic coupling between the material in the sliding track and the unworn material outside the sliding track, is thus not taking place in this case.

The open circuit potential of Fe–31% Ni is not affected by the sliding contact conditions (see Figure 4.5).

Variations of the open circuit potential of materials subjected to sliding conditions can thus be correlated with variations in the surface conditions of the material under

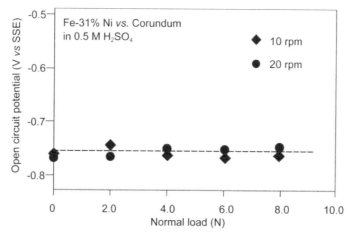

4.5 Variation of the open circuit potential with sliding parameters (normal load, contact frequency) for Fe–31% Ni immersed in 0.5 M H_2SO_4. Counter-body: flat-ended corundum pin. Sliding speed at 10 rpm was 0.0079 ms^{-1}, and at 20 rpm 0.016 ms^{-1} [28]

investigation. A detailed interpretation of the variation of the open circuit potential encounters however a major difficulty. Indeed as noticed above, the open circuit potential is an average value determined by factors as the ratio active-to-passive material in the sliding track, the repassivation kinetics of the base material, the contact frequency, and the normal load. As a consequence, the local surface conditions of the material in and outside the sliding track cannot be precisely derived from open circuit measurements unless more detailed electrochemical data on passive and active material become available. Such information on the local tribological and electrochemical conditions across a partly worn surface could be gained from a precise knowledge of the potential distribution over that surface. Microelectrodes could be helpful in that respect. Some reviews are available showing the use of such microelectrodes to investigate localized corrosion [5] and corrosion–wear [12]. However, the interpretation of the variation of open circuit potential and potential distribution, requires kinetic data to characterize the electrochemical reactions occurring on active and passive materials. Such kinetic data can be acquired by electrochemical techniques such as potentiodynamic polarization and electrochemical impedance measurements.

4.3   Polarization curves

4.3.1   Potentiodynamic polarization measurements

Potentiodynamic polarization measurements can be used to derive the dependence of anodic or cathodic current, I, on the electrode potential, V, measured versus a reference electrode. This method is most useful to determine the active/passive behavior of materials at different potentials.

4.3.2   Application to AISI 316 stainless steel

Potentiodynamic polarization curves obtained at increasing potential dV/dt (direct scan) on AISI 316 stainless steel immersed in 0.5M H_2SO_4 are shown in Figure 4.6.

   Two case studies are shown, namely one without any external loading (dashed line) and the other under loading at 15 N in a sliding contact against a corundum ball (full line).

   The curve obtained without applied load (dashed line) reveals the existence of a wide passivation plateau between −0.5 and +0.5 V/SSE. Below that potential range, hydrogen evolution is observed at cathodic potentials while anodic dissolution takes place at potentials above +0.5 V/SSE. At a potential of −0.3 V/SSE which corresponds to the zero current potential ($I = 0$), the stainless steel is protected from corrosion by a passive film. At this potential value, the measured current, I, is the sum of two partial currents I_{ox} and I_p with I_{ox} the current originating from the reduction of oxygen dissolved in the test solution, and I_p the current originating from the passive film. I_{ox} and I_p are not zero but $I_p = -I_{ox}$

$$I = I_{ox} + I_p \qquad\qquad [4.2]$$

When a normal load (15 N) is applied under sliding, the shape of the polarization curve changes drastically (see full line in Figure 4.6). An anodic current of about 0.5 to 1.0 mA appears in the potential range [−0.7; +0.5] V/SSE, indicating a dissolution of the material. At a potential of around −0.7 V/SSE, corresponding to the zero

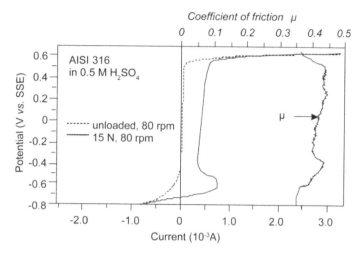

4.6 Potentiodynamic polarization curve on AISI 316 stainless steel immersed in 0.5 M H$_2$SO$_4$. Variation of the coefficient of friction μ during the potentiodynamic scan. Surface area: 4.9 cm^2. Potential scan rate: 1.6 mVs^{-1}. Potential scan from -0.8 to $+0.6$ V/SSE. Sliding speed 0.126 ms^{-1} [28]

current potential ($I = 0$), it is possible to calculate an equation similar to [4.2] in which I is also considered as the sum of two partial currents but in this case I_t and I_p:

$$I = I_t + I_p \qquad\qquad [4.2^{\text{bis}}]$$

where I_t is the current originated from the sliding track area where the passive film is partially destroyed and the material is active, and I_p the current linked to the surface not subjected to sliding that remains in the passive state.

At the zero current potential, $I_p = -I_t$, while a galvanic coupling exists between the sliding track area and the surface not subjected to sliding. These partial currents flow thus between the sliding track area and the rest of the surface. In the sliding track where dissolution of the material and the formation of a new passive film occur, I_t is anodic. On the remaining surface, I_p is cathodic and is related to reactions as the dissolved oxygen reduction or the hydrogen reduction.

When the potential increases above the zero current potential, the conditions of galvanic coupling evolve, and I_t is not anymore equal to $-I_p$. As a result, the current, I, flowing between the specimen and the counter electrode increases. On the surface not subjected to sliding and being in the passive state, I_p cannot exceed the value of the current measured at the same potential on the unloaded specimen. By comparing the values of I under both conditions (see Figure 4.6), it can be deduced that under sliding $I = I_t$.

The current I measured under sliding and its evolution with the applied potential, are characteristic of the behavior of a material in the sliding track. The steep increase of the current with potential around the zero current potential, indicates that a rapid dissolution occurs in the sliding track. The decrease of the dissolution current above -0.6 V/SSE reveals the effect of repassivation in the sliding track. The kinetic of this reaction, increases with potential. This induces a lowering of the total depassivated area, and thus a decrease of the dissolution current.

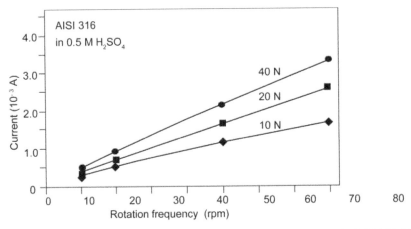

4.7 Variation of the anodic current on AISI 316 steel immersed in 0.5 M H₂SO₄ at an applied potential of 0 V/SSE, with rotation frequency of a corundum pin and for various applied normal loads. Counter-body: corundum pin. Sliding speed 0.016 ms⁻¹ at 10 rpm [28]

Another aspect of the complex interaction between processes governing corrosion and wear can be seen in Figure 4.6 for AISI 316 stainless steel based on the variation of the coefficient of friction, μ, during a potential scan. The coefficient of friction reflects the surface state of a material in the contact area. In Figure 4.6, a high coefficient of friction is noticed at polarization conditions where a large tendency to passivation prevails. Around the zero current potential, the lowest coefficient of friction is noticed what could be to the fact that in this potential range the mechanical depassivation in the contact area is at maximum and thus leads to the largest area of active material.

It must be noted that at potentials between −0.7 and +0.6 V/SSE, the current I increases with increasing rotation rate (Figure 4.7) and thus decreasing exposure time in between successive contact events.

Both effects can be explained by an increase of the area depassivated by friction. Indeed, at increasing sliding speed, the area depassivated per unit of time increases whereas the restoration rate of the passive film remains constant. On the other hand, at increasing normal loads the contact area in which the destruction of the passive film occurs, increases, and as a result the total active area also increases.

4.3.3   Application to Fe–31% Ni

In analogy, the potentiodynamic polarization curves obtained at increasing potential dV/dt (direct scan) on Fe–31% Ni, immersed in 0.5M H₂SO₄ are shown in Figure 4.8.

Two cases are shown, namely without any external loading (dashed line) and under sliding contact conditions against a corundum ball loaded at 75 N (full line).

In both cases, a steep increase of the current is observed around the zero current potential (−0.75 V/SSE). This indicates that the material is in an active state and subjected to corrosion in both tests. This result is in agreement with the open circuit measurements in Figure 4.5. The passivation of Fe–31% Ni occurs only at potentials above +0.1 V/SSE, where a steep drop of the anodic current is noticed [13]. Under

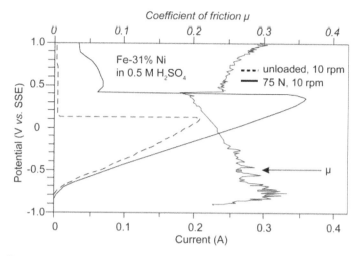

4.8 Potentiodynamic polarization curve obtained on Fe–31% Ni immersed in
0.5 M H_2SO_4. Variation of the coefficient of friction, μ, during potentiodynamic
scans. Surface area: 2.3 cm². Potential scan rate: 1.6 mVs^{-1}. Potential scan from
−1.0 to +0.1 V/SSE. Sliding speed 0.0079 ms^{-1} [28]

sliding (full line), a higher potential is required to form a passive film. The passivation
current remains much higher than the current measured without loading, because
the passive film is continuously damaged and restored in the contact area. The dissolution rate of the metal is also higher in the active potential range under sliding
conditions.

Another aspect of the complex interaction between processes governing corrosion
and wear can be seen in Figure 4.8 for Fe–31% Ni. This figure shows the variation of
the coefficient of friction, μ, during the potential scan. The decrease of μ when the
dissolution rate increases, was related to the appearance of a thin black layer of corrosion products on the surface of the specimen. The decrease of μ can result from a
lubricating effect generated by this layer that is growing continuously and acts as a
third body in the contact. It is important to note (see Figure 4.8) that μ increases
steeply when the current suddenly drops due to passivation. When passivation occurs
during the potential scan, the black layer peels off from the surface of the specimen
and disappears within a few seconds, probably due to dissolution in the electrolyte.
The simultaneous increase of μ and the vanishing of the black layer confirm a
lubricating effect of corrosion products.

The evolution of the dissolution kinetic of Fe–31% Ni at a given potential with
tribological parameters is shown in Figure 4.9.

The variation of the current I measured at −0.3 V/SSE is plotted versus the rotation
rate of the pin and that for different normal loads. An increase of the anodic current
with rotation rate is observed. Such an evolution is observed at any potential in the
active range. This evolution can be explained by the fact that friction wipes off the
layers of adsorbates and corrosion products locally from the contact areas, and
creates what can be called "naked" areas, where dissolution takes place at a higher
rate. After the sweep, layers of adsorbates and corrosion products are restored in
these areas and the dissolution current density decreases. When the rotation and the
frequency increase, the naked area created per unit of time increases, and thus the
total current increases.

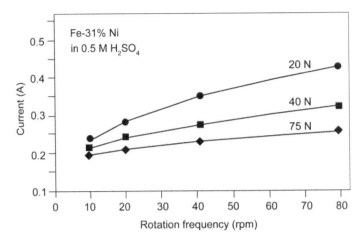

4.9 Variation of the anodic current on Fe–31% Ni alloy immersed in 0.5 M H_2SO_4 measured at a potential of –0.3 V/SSE, with rotation frequency of the pin for various applied normal loads. Counter-body: flat-ended corundum pin. Sliding speed 0.0079 ms^{-1} at 10 rpm [28]

In the case of Fe–31% Ni, the evolution of the anodic current I with applied load at a given rotation speed, is very different from that recorded on most metals in the active state. In the case of AISI 316 stainless steel, an increase of the normal load at a fixed rotation frequency is expected to induce an increase of the real contact area, and consequently the anodic current. In the case of Fe–31% Ni, as shown in Figure 4.9, the current I decreases at every rotation frequency when the normal load increases. This unusual behavior is due to the influence of straining, induced by sliding, on the electrochemical activity of the metal. Indeed, the dissolution rate of the alloy significantly decreases when the alloy undergoes compressive stresses. Under sliding, when the normal load increases, this straining effect prevails over the increase in contact area.

Polarization curves are thus useful in tribocorrosion studies in so far as they give information on changes in electrochemical reaction kinetics induced by sliding, and also on the influence of the surface reactions on the sliding conditions in the contact. The interpretation of polarization curves suffers from the same difficulty as the one encountered in the interpretation of open circuit potential measurements, namely the coexistence of and galvanic coupling of worn and unworn areas exhibiting different electrochemical states. However, by considering carefully the heterogeneous state of tribological surfaces, polarization curves in tribocorrosion studies can yield detailed quantitative information on aspects as mechanical depassivation of worn surfaces, local and overall corrosive wear rates, and the mechanism of mechanical wear under various tribological conditions [7].

4.4   Potential steps

4.4.1   Potential steps experiments

This method consists in recording the current transient generated by a sudden change in the potential applied to a metallic surface from a potential where the metal is active to a potential where it is passive. This method is used to characterize the kinetics of oxidation (dissolution, passivation) of a material in a given environment. Coupled to measurements of mass variations using a quartz crystal microbalance, and surface analysis techniques, it allows to quantify the relative contributions of dissolution and the formation of the passive film during repassivation of several alloys. The individual contributions of the various components of the alloys could also be determined.

4.4.2 Application to repassivation current transients

This method is used to investigate the repassivation kinetics and consists in studying the repassivation current transient induced by applying a potential step towards the passive region onto a metallic surface from which the passive film has been previously removed.

In these experiments the whole sample surface is in contact with the electrolyte and is at first depassivated by a cathodic pre-polarization. Subsequently a potential step is applied in the passive region, and the resulting repassivation current transient is recorded.

An example of such a current density transient obtained on Stellite 6 alloy (55% Co, 28 % Cr, 3% Fe) immersed in a 0.5 M H_2SO_4 solution, is presented in Figure 4.10.

The current density transient shows a linear decay in a log – log scale. After 0.1 s, the current density i varies as t^{-p}, where $-p$ (dashed line in Figure 4.10) is the slope. This slope varies with the tested metal and electrolyte, and in some cases on the applied potential in the passivation region. A possible equation of $i(t)$ can be:

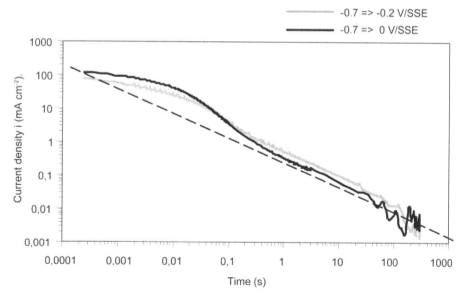

4.10 Current density transients obtained on Stellite 6 in a 0.5 M H_2SO_4 solution, by stepping the potential from −0.7 V/SSE to −0.2 or 0 V/SSE (E_{SSE} = +0.42 V/SHE), in the passivation region [27]

$$i(t) = i_0 \left(1 + \frac{t}{t_0}\right)^{-p}$$

[4.3]

where i_0 and t_0 are constants.

This variation seems to be a general feature of the repassivation transients for most metal – electrolyte systems. This kinetic law is not fully explained by a mechanistic model of passivation until now, but the model of film growth proposed by Jemmely et al. [6], seems to predict the features of the transients and, to some extent, the t^{-p} decay.

It must be emphasized that the characteristics of the transients depend on many parameters:

- The compositions of the metal and the electrolyte,
- The physicochemical conditions (temperature, hydrodynamical conditions in the electrolyte, . . .)
- The electrochemical conditions (applied potential, ohmic drop effects depending on the electrolyte resistivity and geometrical features of the electrochemical cell, . . .)
- The mechanisms of electrochemical reactions, and of passive film growth.

Some of these issues were analyzed in [6]. Actually, this electrochemical approach seems to provide quite similar results as the experiments where an active surface is generated by mechanical activation [7] or even during pitting corrosion, or stress corrosion cracking processes [14] [15].

However, for using current transients obtained in potential steps experiments to explain and to model the results of tribocorrosion experiments, the effects of the different experimental conditions in both experiments must be carefully considered as:

- Geometrical features of the systems under test, resulting ohmic drop effects and potential and current density distributions on the sample
- Potential conditions: constant applied potential, mixed open circuit potential due to galvanic coupling, mixed open circuit potential varying with time, . . .

In tribocorrosion experiments carried out under continuous sliding in open circuit potential conditions, the surface is in a heterogeneous state, the sliding track being partially active contrary to the area out of the track. The open circuit potential conditions are governed by galvanic coupling effects. The resulting potential and current density distributions on the surface are very different from those obtained in potential steps experiments. Experimentally, the local current transients could not be directly measured until now. Nevertheless, the evolution of the corrosive component of wear with the contact frequency was studied and showed that, even in such conditions of galvanic coupling, this evolution could be explained by a repassivation transient corresponding to equation [4.3].

However, it must be noticed that in tribocorrosion experiments performed under continuous sliding, the value of the exponent p was found significantly different from the value given for the same metal – electrolyte system by potential steps experiments. This statement gives a limit to the use of potential steps experiments applied to tribocorrosion studies.

4.4.3   Application to tribocorrosion

The potential step used to generate polarization-like curves has been proposed to quantify the active sliding track area on a passivating alloy subjected to continuous sliding in a pin – on disc tribometer [7]. The active sliding track area represents that part of the sliding track that looses temporarily its passive state due to the mechanical interaction in the sliding area.

The polarization-like curve is derived from polarization pulse tests applied to the samples in absence of any mechanical contact. The surface of the sample is first electrochemically activated by applying a cathodic potential to remove the passive film. Then an anodic potential pulse is applied up to a final potential in the passivation domain. The pulses are applied during a time equal to the rotation period of the pin which is used in the tests under sliding. The resulting repassivation current transient is recorded and is supposed to be similar to the current transient obtained on the sliding track under sliding, the only difference being the area of bare metal from which the current is flowing, namely the total area of the sample in the pulse test, and the active sliding track area where sliding has removed the passive film, in the sliding test. The current density transient is deduced and an average value of the current density is calculated. This procedure is repeated for different pulse potentials in the passivation range of the material. By comparing this curve related to a known active area with the polarization curve plotted in the same potential range under sliding, the active sliding track area A_{act} can be calculated.

4.5   Electrochemical impedance measurements (EIS)

The Electrochemical Impedance Spectroscopy (EIS) is a most performing method for a detailed analysis of electrochemical reactions mechanisms and kinetics. The measurement of the electrochemical impedance is made by superposing a sinusoidal voltage signal with a very small amplitude (5–10 mV) to the open circuit potential. This has the advantage of generating only a negligible perturbation of the electrochemical steady state of the tested material. Impedance diagrams give data on the elementary steps occurring in an electrochemical reaction and on their kinetics. The analysis of these diagrams allows a thorough study of the role of intermediate species adsorbed on the surface and of reaction mechanisms, as well as a study of the properties of passive films. EIS measurements can be implemented only under stationary electrochemical conditions, but theoretical considerations and many experimental studies [16–17], have shown that such conditions can be fulfilled in tribocorrosion tests carried out with a pin-on-disc tribometer if continuous sliding is applied on the surface of the tested material.

Tribocorrosion occurs in industrial field conditions essentially at open circuit potential [18]. Under such conditions, the total current is zero and its anodic component, corresponding to a corrosion or a passivation current, can not be determined by direct measurement. One possible solution, is to use electrochemical impedance spectroscopy (EIS) to measure the polarization resistance R_p at the open circuit potential and to deduce the anodic current from the R_p value.

Many previous corrosion studies have shown that an approximate value of the corrosion current I_{corr} or passivation I_{pass} current of a metal or alloy can be derived from the value of the so-called polarization resistance,R_p, which can be deduced from the impedance diagram measured on a material held either at DC potential corresponding to the stationary open circuit potential value or galvanostatically, zero current. Different methods are available to measure R_p [16]. Among them, (EIS) is the most accurate one.

The electrochemical impedance of an electrode is generally represented by the equivalent electrical circuit given in Figure 4.11.

In this circuit, R_s is the resistance of the solution, CPE is a "Constant Phase Element" accounting for the capacitive properties of the electrochemical double layer, and Z_f is the faradic impedance. R_p, is defined as the limit of Z_f when the frequency tends to zero:

$$R_p = \lim_{f \to 0} Z \qquad\qquad [4.4]$$

The following relationship, known as the second Stern and Geary relation, between the polarization resistance and the corrosion current I_{corr} or the passivation current I_{pass} is generally accepted:

$$I_{corr}\ (or\ I_{pass}) = \frac{B}{R_p} \qquad\qquad [4.5]$$

With B a constant whose value depends on the corrosion or passivation mechanism. The product $R_p\ A_0$ is the specific polarization resistance referred to as r_p.

Several issues were discussed in the literature regarding the validity of such a relationship. They are not detailed here, but it is worthwhile to remind some results:

- The validity of such a linear relation between the current and the inverse of the polarization resistance can only be demonstrated rigorously for a few types of uniform corrosion or passivation mechanisms.
- In the case of localized corrosion, it is not possible to calculate an equation similar to the relationship [4.5] without admitting some simplifying assumptions regarding the mechanisms of anodic and cathodic reactions and the conditions of galvanic coupling between anodic and cathodic areas. The situation of an active sliding track, undergoing anodic dissolution and repassivation drawn by galvanic coupling with macroscopically distinct cathodic region of unworn surface, may rise the same difficulty. Numerical simulations performed on a ball-on-disc geometry indicate that a correct value of the galvanic corrosion current between anode and cathode can be deduced from Rp, corrected of ohmic series resistance, even in fairly resistive media. The condition is to apply the small polarization signal for measuring Rp from a sufficiently distant counter electrode.

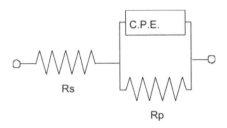

4.11 Equivalent electrical circuit

-   In fact, some studies of localized corrosion allowed to establish empirically the validity of a linear relationship between I_{corr} and the inverse of the measured polarization resistance, even in conditions that did not meet the simplifying assumptions mentioned above [19].

It can therefore be considered using the polarization resistance in a tribocorrosion test procedure to assess the passivation current of a material not subject to sliding, and the corrosion current under galvanic coupling conditions met under sliding.

This method also has the advantage of providing information on the mechanism of corrosion or passivation, which indicates whether a reasonable estimate of the anodic current (corrosion or passivation) can be obtained by applying the relationship 4.5. In particular, when the impedance spectrum measured in a frequency range wide enough (typically from 10^4 to 10^{-4} Hz) reveals the existence of only one relaxation phenomenon, the existence of a linear relationship between I_{corr} and R_p^{-1} can be accepted. This situation arises, for example, when the impedance diagram in Nyquist representation consists of a single arc of circle throughout the frequency domain [20].

Another issue is the value of B. For systems whose mechanism of corrosion or passivation yields relation 4.5, the theoretical calculation shows that B is a function of exponents related to the Tafel rate constants of elementary steps of anodic and cathodic reactions. A rigorous calculation of the value of B is possible only if the overall reaction mechanism is precisely known, and if the values of the exponents are too. Unfortunately, the conditions for such a theoretical approach are not met in most cases of corrosion or passivation of metals and alloys.

However the order of magnitude of B can be specified by calculating the values of B for simplified mechanisms of uniform passivation or corrosion.

For example, let us consider a mechanism of passivation of a metal having the following characteristics:

-   In the vicinity of the open circuit potential, the steady state anodic component of the current (passivation current) is independent of the potential.
-   The cathodic reaction is hydrogen evolution or dissolved oxygen reduction. The rate of the cathodic reaction is not limited by diffusion. This condition is satisfied for oxygen reduction if the anodic passivation current is low enough compared to the limiting current density of diffusion corresponding to the reduction of oxygen dissolved in the electrolyte.

For an active metal M subjected to corrosion, the simplified calculation can be carried out with the following assumptions:

-   The anodic dissolution of M takes place in a single step of charge transfer and its rate is not limited by mass transport.
-   The cathodic reaction is either hydrogen evolution or dissolved oxygen reduction. In both cases, charge transfer is assumed to occur in one single step. The rate of hydrogen evolution is under control of activation kinetics. The rate of the oxygen reduction reaction may be either fully limited by mass transport or only by activation. The case of mixed control is excluded.

For all anodic and cathodic reactions, the values of the transfer coefficients α_a (anodic) and α_c (cathodic) are taken equal to 0.5.

The value of B is then given by the following equation:

$$B = \frac{RT}{F}\left(\frac{1}{n_a \gamma_a \alpha_a + n_c \gamma_c \alpha_c}\right)$$ [4.6]

In this equation, α_a and α_c are the transfer coefficients of the anodic and cathodic reaction, with $\alpha_a = \alpha_c = 0.5$, as above mentioned. n_a and n_c are the number of electrons transferred in the anodic and cathodic reactions respectively:

$$M \rightarrow M^{n_a+} + n_a e^-$$

$$\tfrac{1}{2}O_2 + H_2O + 2e^- \rightarrow 2OH^- \qquad n_c = 2$$

$$H^+ + e^- \rightarrow \tfrac{1}{2}H_2 \qquad n_c = 1$$

The values of γ_a and γ_c are:

– $\gamma_a = 0$ for passivation.
– $\gamma_a = 1$ for corrosion.
– $\gamma_c = 1$ if the cathodic reaction is hydrogen evolution.
– $\gamma_c = 1$ if oxygen reduction is the cathodic reaction and its kinetics is under control of activation.
– $\gamma_c = 0$ if oxygen reduction is the cathodic reaction ant its kinetics is under control of mass transport.

For uniform corrosion, values are gathered in Table 1:

Table 4.1   Values of the coefficient B calculated for uniform corrosion with various values of n_a and n_c

γ_c	n_c	n_a 1	2	3	4
1	1	26 mV	17 mV	13 mV	10 mV
1	2	17 mV	13 mV	10 mV	8.5 mV
0	2	52 mV	26 mV	17 mV	13 mV

For passivation, the following values of B are derived from equation [4.6]:

$B = 26$ mV if the cathodic reaction is oxygen reduction.
$B = 52$ mV if the cathodic reaction is hydrogen evolution.

For an alloy, the rigorous calculation of B becomes extremely complex. An approximate value can be calculated by equation 4.6, taking for n_a the average value calculated from the values of n corresponding to the metallic constituents of the alloy and from their relative proportions in the alloy.

The passivation current can be determined at the beginning of a tribocorrosion test before applying sliding, from the value of R_p measured by EIS, when the sample is in a uniform and stationary electrochemical state.

Under continuous sliding, a value of the anodic current I_{act} flowing from the active area (sliding track) to the surrounding passive areas can be calculated from the value of R_p deduced from the impedance diagram measured under sliding.

The following two examples illustrate the usefulness of this method for the study of tribocorrosion.

4.5.1   An application to AISI 316 stainless steel

Impedance diagrams were recorded on the AISI 316 disk during the experiment shown in Figure 4.2. The impedance measurements were carried out during the time intervals marked 1 to 4, corresponding to the material behavior before sliding, namely the time intervals marked 1 and 2, during sliding, being time interval 3, after sliding, namely time interval 4. The impedance diagrams are represented in a Nyquist plot in Figure 4.12.

Figure 4.12a shows two impedance diagrams with the polarization resistance R_p values:

– R_{pol1} for time interval 1 where the material is in a completely active surface state,
– R_{pol3} for time interval 3 where the material is locally depassivated by sliding.

Figure 4.12b shows two impedance diagrams with the polarization resistance, R_p values:

– R_{pol2} for time interval 2 where the material is in a completely passive surface state,

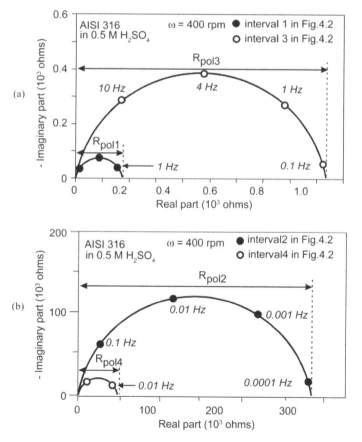

4.12 Nyquist impedance diagrams recorded at open circuit potential during the experiment of Fig. 4.2 [28]

    – R_{pol4} from time interval 4 where the material is locally repassivated after sliding stop.

For both Nyquist plots, the single semi-circle can be explained by a double layer capacity coupled to the charge transfer due to electrochemical reactions (corrosion, passivation, etc.). In the case of the Nyquist plots in Figure 4.12, the polarization resistance is directly related to the size of the semi-circle. It is interesting to compare the impedance diagrams corresponding to a fully active surface state (Figure 4.12a – R_{pol1}), with the one of a totally passive surface state (Figure 4.12b – R_{pol2}). The size of these impedance diagrams differs by several orders of magnitude, particularly in the "low frequency range" (typically < 0.1 Hz).

- Figure 4.12a – R_{pol1} recorded at the beginning of the test before passivation of the alloy corresponds to a totally active surface. According to equation [4.5], the corrosion current is:

$$I_{act1} \text{ is } 0.14 \times 10^{-3} \text{ A (with } B = 26 \text{ mV)}$$

Considering the total area of the specimen (4.9 cm²), the corrosion current density is:

$$i_{act1} \text{ is } 0.028 \times 10^{-3} \text{ A cm}^{-2}$$

- Figure 4.12b – R_{pol2} recorded after passivation and before applying any sliding corresponds to a fully passivated surface. The residual corrosion current (passivation current) can be assessed in equation [4.5] from the value of R_p determined in the impedance diagram ($R_p = 280 \times 10^3 \Omega$)) so that:

$$I_{pass} = 89 \times 10^{-9} \text{ A (with } B = 26 \text{ mV)}$$

corresponding to a corrosion current density:

$$i_{pass} \text{ of } 18 \times 10^{-9} \text{ A cm}^{-2}.$$

Despite the quite similar open circuit potentials in time interval 1 before sliding and time interval 3 during sliding (Figure 4.12a), and on the other hand in time interval 2 before sliding and time interval 4 after sliding (Figure 4.12b), the impedance diagrams show important differences:

- In the case of Figure 4.12a, the material is either completely active (corresponding to the time interval 1 in Figure 4.2) or locally depassivated by sliding (corresponding to the time interval 3 in Figure 4.2). A clear difference is observed on comparing these diagrams, whereas the value of the open circuit potential does not allow any distinction.
- The Figure 4.12b is related to a passive surface before sliding (corresponding to the time interval 2 in Figure 4.2) and after sliding (corresponding to the time interval 4 in Figure 4.2), respectively. The important difference in the size of the impedance diagrams reveals again that, despite very similar open circuit potentials, the electrochemical state of the material cannot be considered as identical in both cases.

These differences can be interpreted by a simple model based on an equivalent electrical circuit of the impedance of the specimen undergoing the tribocorrosion test shown in Figure 4.13. In a first approach, the global impedance of the specimen is considered as a parallel connection of impedances Z_a and Z_p. Z_a represents the impedance of the

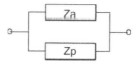

4.13 Equivalent electrical circuit representing the impedance of the specimen surface by a parallel combination of the impedance of the active and passive areas (Z_a and Z_p, respectively).

material in an active electrochemical state where the passive film was destroyed e.g. by friction, and Z_p the impedance of the surface still covered by a passive film.

For such a model, the global impedance is given by:

$$1/Z = 1/Z_a + 1/Z_p \qquad [4.7]$$

It must be noticed that this relation is valid over the whole frequency range, and in particular at very low frequencies where $Z \cong R_p$. Consequently, a relation similar to equation [4.4] can be derived for the polarization resistance R_p, R_{act}, and R_{pass} (polarization resistances relative to the total area, the active area and the passive area, respectively).

The value of R_{act} is deduced from the following formula:

$$\frac{1}{R_{act}} = \frac{1}{R_p} - \frac{1}{R_{pass}} \qquad [4.8]$$

- In the case of the Nyquist diagram corresponding to time interval 3 (see Figure 4.12a), it may be assumed that, under sliding, in the low frequency range, the impedance Z_p of the passive area of the specimen remains much greater than the impedance Z_a of the area depassivated by sliding. In that case, the total impedance Z and the impedance Z_a should have very close values as well as the polarization resistances R_p and R_{act}. This means that the low frequency part of this diagram can represent in fact the impedance Z_a of the active area depassivated by sliding in the sliding track. From equation [4.3], the corrosion current, I_{act} can be assessed as:

$$I_{act} = 27 \times 10^{-6}\ \text{A (with } B = 26\ \text{mV)}.$$

If one assumes that the electrochemical activity of the area depassivated by friction is the same as the one of the active surface before sliding, the value of the current density i_{act1} can be used to evaluate the area A_{act}:

$$A_{act} = I_{act} / i_{act1} = 1\ \text{cm}^2$$

Owing to the value of the sliding track area (0.5 cm²) measured in this sliding test, the value of A_{act} appears overestimated. Different reasons related to the simplifying assumptions made for the interpretation of the impedance diagrams, can explain this:
- The parallel combination of Z_a and Z_p can be considered as a valid model only in a limited frequency range and for particular conditions concerning the arrangement of the electrodes, the geometry of the cell, of the specimen, and of the active and passive areas on the specimen. In our test conditions, the model can be considered as valid in the low frequency range.

- The assumption $Z_p \gg Z_a$ should be verified, in particular at the open circuit potential corresponding to test conditions. This can be achieved by measuring the impedance of a passive specimen polarized at the suitable potential value. In the used test conditions, such an impedance measurement demonstrated indeed that the assumption is valid.
- The assumption that the active areas have the same activity as the specimen before passivation (same current density), is the most contestable one. The evaluation of i_{act} and A_{act} requires a more detailed investigation of the evolution of the tribocorrosion mechanism and the wear kinetics with the tribological conditions [7].

Impedance measurements provide also interesting information on the state of AISI 316 stainless steel after unloading corresponding to time interval 4 (Figure 4.12b). When the open circuit potential returns close to the value of the passive state before sliding. The different size of diagrams corresponding to time intervals 2 and 4 (Figure 4.12b) reveals that, after sliding, despite the high value of the open circuit potential, the initial passive state of the surface was not restored. The smaller size of diagram 4 ($R_{pol4} < R_{pol2}$) indicates that, in the sliding track, the passive film has not yet regained its initial thickness and protective properties.

4.5.2   Application to Fe–31% Ni

The Nyquist impedance diagrams on Fe–31% Ni are presented in Figure 4.14. These diagrams were recorded on Fe–31% Ni immersed in 0.5M H_2SO_4, at an anodic potential of –675 mV/SSE (+100 mV/open circuit potential), before sliding (unloaded diagram), and under sliding. At this potential, the prevailing reaction is dissolution.

For the unloaded diagram, before sliding, the EIS diagram consists of two semi-circles:

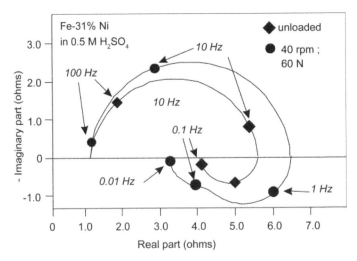

4.14 Nyquist electrochemical impedance plots recorded at $E = -675$ mV/SSE ($I = 20$ mA). Fe–31% Ni immersed in 0.5 M H_2SO_4. Surface of the sample = 2.3 cm². Counter-body: flat-ended corundum pin. Sliding speed 0.031 ms⁻¹ [28]

- The capacitive one, at highest frequencies, shows the influence of the dielectric properties of the electrochemical double layer, and of the charge transfer through this double layer due to electrochemical reactions.
- The inductive arc, at lowest frequencies, can be explained by the relaxation of the surface concentration (sinusoidal variation with a phase shift against measurement of the voltage signal) of an adsorbed intermediate species involved in the dissolution mechanism. Under sliding, the size and the shape of the diagram change:
- The size of the "high frequency" capacitive arc increases, indicating an increase in the transfer resistance. This can be interpreted by a decrease in the "reactivity" of the surface. This result is consistent with the straining effect pointed out in polarization measurements.
- A second inductive arc appears in the impedance plot revealing the presence of at least two adsorbed intermediates in the dissolution mechanism. Before sliding, at −675 mV/SSE, only one of them gives rise to an inductive loop. The surface concentration of the other adsorbate does not vary enough with voltage to have a significant contribution in the impedance.

Under sliding, the kinetics of the dissolution process is apparently modified, as revealed by the second inductive loop in the diagram. A detailed investigation of these diagrams permits a further unraveling of the kinetics of the adsorption of these intermediates on the surface, as shown in literature [21].

These impedance measurements carried out during tribocorrosion tests illustrate the usefulness of this method to study the mechanism of electrochemical reactions involved in tribocorrosion processes, and the interaction between corrosion and friction. More detailed information on tribocorrosion processes are expected from a systematic study of impedance diagrams recorded at varying tribological test conditions (variation of the normal force, sliding speed, rotation frequency, etc.). The interpretation of impedance measurements recorded during sliding tests is difficult due to the heterogeneous state of the surface, as in the case of polarization measurements. In fact, a non-uniform distribution of the electrochemical impedance over the disk surface must be considered. The action of friction can be thoroughly analyzed, only if this distribution is known. In tribocorrosion experiments, a local analysis of the electrochemical state is thus necessary to interpret impedance measurements. Research work on electrochemical systems with a non-uniform distribution of the electrochemical impedance is available now [22]. Such measurements and models can be adapted to tribocorrosion conditions. In that field, microelectrodes could help to map the electrochemical impedance on disk surfaces. Electrical equivalent circuit models or finite element models could be used to get distributions of impedance, and to calculate the overall impedance.

4.6   Electrochemical noise analysis (EN)

The use of electrochemical noise (EN) measurements at OCP for the investigation and monitoring of tribocorrosion has allowed many interesting advances in the corrosion science in recent years [23–26]. A special advantage of EN measurements includes the possibility to detect and study the early stages of localized corrosion and likely to isolate individual events related to the film breakdown in the sliding zone. Under sliding, the sliding track is considered as a part of the surface undergoing localized corrosion. The area outside the sliding area is considered in the passive state surface. Nevertheless, the understanding of the electrochemical information included in the EN signal is actually limited. The main difficulty is the role of the cathodic processes on the EN signals, in particular the reduction of dissolved oxygen, which remains uncertain and has not been sufficiently investigated until now.

4.7   Multi-scale approaches in Tribocorrosion

Damage in tribocorrosion results from the combined action of corrosion and mechanical loading. It depends on the properties of the contacting materials (the microstructure, for example), the mechanics of the tribological contact (the mechanical stress/strain field at the specimen surface, for example) and the physical-chemical properties (corrosion). Therefore studying processes in tribocorrosion requires an integration of researchers and engineers from various disciplines, namely, material science, mechanical engineering, surface engineering, electrochemistry/chemistry, tribology, biology, and medicine.

4.7.1   Classical approaches at the macro-scale

Advantages and limitations of global approaches: The most important step in tribocorrosion analysis of a system is to understand the combined as well as individual effects of tribology and corrosion along with their interaction which can either be beneficial or detrimental. In a first approach, classical methods consisting in measuring at the macro-scale the coefficient of friction and some electrochemical parameters can be very useful [28–33].

However, classical techniques are limited to surface-averaged measurements which account for the behavior of the whole material surface. In tribocorrosion problems of heterogeneous surfaces, the response associated with local phenomena (pitting, breakdown of the passive film, micro-cracking, repassivation. . .) cannot be extracted from the global measurements. To overcome these difficulties, several scanning techniques have been developed.

Challenges for developing multi-disciplinary approaches at micro-scale: The objective of multi-disciplinary approaches in tribocorrosion is to monitor in-situ mechanical (strain components, plasticity. . .) and electrochemical parameters (pH, potential, species concentrations) simultaneously and to quantify possible synergistic effects. This is an important challenge to researchers because synergistic and antagonistic effects between mechanical and chemical mechanisms strongly affect the rate of degradation. This must be done under strictly controlled mechanical conditions, using materials with well-characterized surface properties. In addition, because of the complexity of the surfaces generated during tribocorrosion (including corrosion products, pits, micro-cracks, stress concentrations, scratches. . .), local approaches have to be used to determine the acting/driving mechanisms in association with material loss as a function of the selected parameters [34–36]. At the micro-scale, the single reaction steps, such as local activation and repassivation processes, can be studied individually. Despite numerous studies on this subject direct experimental evidence of these local processes is still missing. This is an important scientific challenge to gain new insights into passivity and its local breakdown.

At this point, there is a technical challenge. Classical tribometers (commercially available or homemade) have to be modified to incorporate electrochemical and mechanical microprobes. Some of the limitations of a modified system can be listed as follows [37];

(i)   geometry and construction of a corrosion cell (appropriate for the tribometer),
(ii)   proper/consistent locations of electrode,
(iii) possible leakage of the solution/electrolyte, and
(iv)   collection and synchronization of the data from the tribometer and potentiostat.

In addition, this multiprobe configuration can be prone to artifacts due to the presence of several tips in the aqueous solution in the close vicinity of the contact area and markers at the specimen around the contact area. Perturbations of potential/current lines and species concentrations have to be first investigated on a well-charaterized system.

In this context, scientific challenges consist in characterizing mechanical and electrochemical properties of passive films and their galvanic coupling with sliding track, monitoring the evolution of pH, species concentrations at the local and to correlate evolutions with microstructural/mechanical changes... Another scientific challenge is to bring together the fundamental understanding resulting from multi-disciplinary approaches at the local scale into predictive models. These models must incorporate mechanical-(electro)chemical effects on the formation of debris, material evolution, active dissolution and repassivation.

4.7.2   Local techniques for characterizing surface modifications due to tribocorrosion, and related multi-disciplinary approaches

Local techniques in mechanics, (electro)chemistry and surface analysis, which are commonly used in tribocorrosion or which are relevant to this field, are described in this section.

Characterization methods in surface mechanics: Tribocorrosion may generate relatively complex 3D strain fields in materials (compression, tension, shear...). These strain fields depend on the material properties and microstructure, the formation of an oxide film or corrosion products, the propagation of micro-cracks, and the solicitation mode. In the case of heterogeneous microstructures, large strain concentrations may be locally observed. On the other hand, frictional shear strain may generate severe plastic deformation and high density of dislocations.

As it was already mentioned, the presence of this complex 3D strain field affects significantly the electrochemical behavior of metallic alloys (dissolution kinetics, repassivation mechanisms...). It is therefore important to quantify these mechanical parameters. In order to identify the most appropriate experimental techniques to be used, different issues have to be addressed.

– **what is the scale of measurements?** This mainly depends on the specimen microstructure and the activation mode.
– **what space-time resolution is required to capture non-uniform full-field deformations and to provide sufficiently accurate data?** Real-time in-situ measurements generally require the use of both spatially- and time-resolved techniques. A compromise between spatial and temporal resolution must generally be found. For example, high spatial resolution requires long acquisition times to acquire a complete image.

In addition, the selected techniques must have the following features: no contact with the specimen required, compatible with the geometry of a tribometer, and a direct measurement that does not require recourse to a numerical or analytical model.

In the recent past, thanks to the sharp advances in microcomputer and camera technology, many research groups devoted to optics, experimental mechanics or data processing have been developing suitable techniques based on the use of optical devices, digital cameras, algorithms and softwares which automatically process images.

Several types of full-field techniques have been proposed in the past years [38]. These techniques can directly provide displacement or strain contours onto specimens under tribocorrosion testing (around the sliding track). The nature of the measurements can be displacement, strain or temperature. Displacements are measured with various techniques [39], for instance speckle [40], speckle interferometry [41], geometric Moiré [42], Moiré interferometry [43], holographic interferometry [44], image correlation [45] or grid method [46]. Strain can be obtained by numerical differentiation of the above displacement fields with suitable algorithms [47] or directly, for instance with shearography [48–49], speckle shearing photography [50] or by Moiré fringes shifting [51].

Regarding image correlation, the preparation of the surface is very simple and the displacements are easily obtained by matching different zones of two images captured before and after loading of the specimen. Surfaces under investigation can therefore be prepared with white painting and sprayed with a black aerosol. This leads to a random structured aspect, which can be observed with a digital camera. Lithography can also be used to deposit regular arrays of dots [52], Figure 4.15. In this case, the diameter of dots and the dimension of the arrays can be chosen according the specimen microstructure and the applied load. A review of the literature indicates that these techniques have never been applied to tribocorrosion problems. Other techniques, such as the X-ray diffraction technique coupled with different analysis

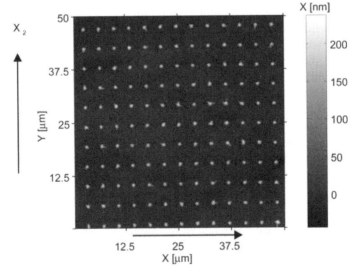

4.15 AFM image AFM of an array of gold pads deposited at the surface of stainless steel polished using 1 µm diamond paste (pad diameter: 0.4 µm; distance between pads: 3.6 ± 0.02 µm along X_1 and 4 ± 0.02 µm along X_2) [51]

methods can be used to measure ex-situ (after tribocorrosion test) the surface stress field [53].

Finite-element methods also appear as an attractive way of evaluating the distribution of plastic strain in the contact region. They are expected to permit isolation of the individual effects of operating parameters such as load, sliding speed and sliding distance on the wear process. . . The F.E simulation can be coupled with a wear algorithm based on the Holm-Archard equation to predict the development of an inclination in the wearing profile of the pin due to uneven pressure distribution at the contact zone [54]. In these methods, particular attention has to be paid to the local contact area, strain-stress relationships, material models and boundary conditions.

Electrochemical methods: During the past few years various local electrochemical techniques have been developed which allow local measurements. Some of these techniques have already been combined with a tribometer to perform local in-situ measurements. The other techniques can be adapted to be inserted in a tribological system. Two fundamentally different methods are applied for local investigations:

- **Scanning techniques**, where the immersed metal surface is scanned with a microprobe to measure the potential in a static mode (Scanning Reference Electrode Technique SRET) or the current in a vibrating mode (Scanning Vibrating Electrode Technique SVET). A conventional SRET has been modified to provide a novel method for characterizing the real-time localized tribocorrosion behavior of uncoated and PVD C/Cr coated samples during rubbing [55]. The influences of various testing conditions involving contact load and test environments have been demonstrated, and have been found to affect some electrochemical parameters (local anodic current, number of active anodic areas, sample (repassivation) recovery time, free corrosion potential) and mechanical parameters (amount of wear, wear coefficient and wear morphology). More recently, the Scanning electrochemical Microscope (SECM, Figure 4.16) allows the mapping of species distribution around an active site.

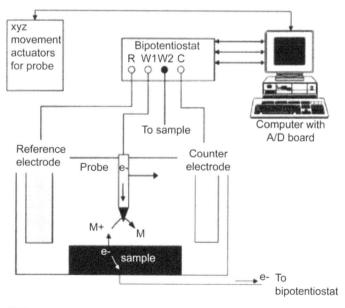

4.16 SECM Block Diagram [56]

The SECM probe is an ultramicroelectrode, which is an electrode of nanometer to micrometer dimension. This microscope can be used in the generation or collection modes. The ultramicroelectrode can work as an amperometric sensor of chemical species in solution, allowing the measurement of local differences in electrochemical reactivity on the scanned substrate. The SECM has been applied to corrosion research in various occasions [56]. It has not yet been applied to tribocorrosion problems.

Several types of microprobes can also be used to measure locally the evolution of pH and potential around the active sites and the sliding track. Concerning the mapping of potential, silver-silver chloride microelectrodes can be made by using pulled microcapillaires with a tip diameter in the range of 10–100 μm. Regarding the evolution of pH close to the surface, a quinhydrone electrode, which is based on the electrochemically reversible oxidation-reduction system of p-benzoquinone and hydroquinone in which hydrogen ions participate, can be considred. The construction of this microprobe is as follows : a platinum wire is introduced into a microcapillary (tip diameter between 10 and 100 μm) which is filled with a solution containing some crystals of quinhydrone. The potential of this microprobe depends on the pH of the adjacent solution. Local pH measurements can also be performed using tungsten wire insulated with polyimide. The microprobe tip must be freshly cut. Such microprobes can be fixed on X-Y-Z motors and surface mapping is carried out above and around heterogeneities under potentiostatic control and at the open circuit potential on both samples.

– **Methods where only small areas are exposed to the electrolyte**: Figure 4.17a shows an electrochemical microcell technique using either micro-capillaries (diameter in the range of 50 nm to 1 mm, or by covering the residual surface with lacquers or photoresists [57].

In the latter case, a layer of silicon rubber is deposited at the microcapillary tip (Figure 4.17b). This ensures a good contact between the microcapillary and the specimen surface, no crevice corrosion and no leakage during measurements. The

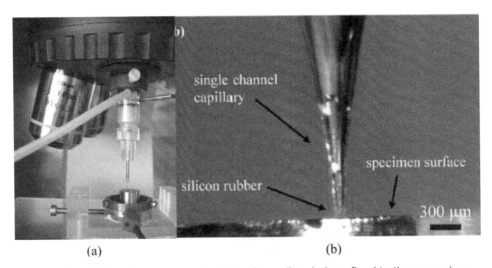

(a)   (b)

4.17 (a) View of the electrochemical microcell technique fixed to the nose piece of a metallurgical microscope. (b) View of the contact between a micro-capillary and a specimen surface [57]

influence of the cell geometry on the electrochemical response can be determined using numerical simulation [58–59]. Based on microcapillary techniques, a new method [60] to study tribocorrosion properties has been developed, where measurements under controlled electrochemical and mechanical conditions can be performed. The micro-electrochemical tribocorrosion technique allows one to study the repassivation kinetics of surface areas activated by tribocorrosion processes. A detailed description of the method is given in Ref. [61]. The rubbing partner (an Al_2O_3 tube) rotates inside the microcapillary. Rotational speed, load, and applied potential or current can be varied in-situ.

4.8 Conclusions

Various electrochemical methods (open circuit potential measurements, polarization curves, EIS measurements) performed during tribocorrosion tests, can provide essential information on the tribocorrosion mechanism, and also on the kinetics of these reactions that control the corrosion component of the total wear loss. Aspects of tribocorrosion mechanism that can be investigated are the nature of electrochemical reactions, existence of protective passive films, interactions between electrochemical reactions and friction. Related to the kinetics, information can be gained on corrosion rate, rate of depassivation by mechanical action in the contact area, and rate of passive film restoration.

In addition, electrochemical measurements can give information on the tribological conditions of friction and wear mechanisms, because the electrochemical state of the surfaces under sliding is highly dependent on these conditions, and also on the wear process in the contact (mild oxidation, or abrasive wear, etc.). This aspect was recently developed in tribocorrosion studies [7].

The use of electrochemical methods in tribocorrosion is still facing a few limitations, related to the particular conditions of tribocorrosion tests. These limitations are mainly due to the heterogeneous state of the surface subjected to sliding, and the time evolution of the rubbed surfaces resulting from the wear process:

- In most tribocorrosion tests, in particular those carried out with pin-on-disk apparatus, only a fraction of the specimen surface is undergoing sliding. If both sliding track and the unworn surface are in contact with the corrosive electrolyte, a galvanic cell (corrosion cell) between the rubbed areas and the rest of the surface, must be considered. The galvanic cell induces an electrochemical potential distribution over the surface of the specimen, which can control to a large extent the rate of electrochemical reactions in the sliding track and on the rest of the surface. This effect can even, in particular circumstances, induce radical changes in the electrochemical state of this latter part of the surface, by destabilizing a protective passive film. In such an extreme case, friction applied on a limited fraction of the surface of a passive material may induce a general corrosion over the entire surface [62]. In tribocorrosion, the heterogeneity of the specimen surface and the resulting effect of galvanic cell must be considered when selecting an electrochemical method, the measurement conditions, and the models for interpretation. In that respect, localized corrosion studies (pitting corrosion, crevice corrosion, stress corrosion cracking) and the methodological and theoretical tools developed for the study of this category of corrosion processes, should be very useful. For example, the interpretation in this paper of the impedance data of AISI 316 steel, was based on a very simple equivalent circuit. This model, involving a parallel combination of impedances of active and passive areas, was already proposed in localized corrosion studies. A more detailed and quantitative interpretation of the diagrams requires a model in which the effect of the distribution of potential and current generated by galvanic coupling is included.

- The evolution of the characteristics of the sliding track (width, roughness, mechanical properties, etc.), the contact area, and the local tribological conditions during the wear process, are also parameters to be considered when selecting electrochemical measurements conditions, and interpretating data. In a pin-on-disk tribometer, a circular sliding track is generated on the disk surface through a continuous sliding movement (uni-directional sliding), at a constant sliding speed. In that case, pseudo-stationary electrochemical conditions are achieved, and this allows to implement various electrochemical methods as polarization and impedance measurements, which require stationary states during the measurements. However, if global pseudo-stationary conditions can be obtained during a limited time, it must be considered that, at the local scale on every point of the sliding track, transient electrochemical conditions are encountered with periodic activations. In a thorough investigation of the tribocorrosion process, the measurement procedures and the theoretical models must take into account such local transient and periodic aspects. In addition, an evolution of the global stationary state must be considered as the result of changes in sliding track features and local tribological conditions. In some sliding tests, like pin-on-flat in which the pin undergoes an alternating movement (bi-directional sliding), pseudo-stationary states are not obtained along the whole sliding track. As a result, polarization or impedance measurements cannot be implemented. On the other hand, in such tests, the analysis of open circuit potential transients, or current transients obtained under potentiostatic conditions, provides information on the degradation of the passive film and on the kinetics of the restoration of the passive film. A full understanding of the tribocorrosion mechanisms requires finally and in-depth correlation of electrochemical data and surface characterization data.

Significant progress on the study of tribocorrosion mechanisms requires some technical advances (to combine tribometer with microelectrodes in mechanics and (electro)-chemistry). This will open new routes to quantify the galvanic coupling between the sliding track and the surrounding surface, to monitor electrochemical parameters such as pH and potential at the micro-scale, and to correlate these parameters to microstructural evolutions. Therefore particular attention has to be paid to the development of novel tribometer systems including microsensors.

Some information obtained at the micro-scale have then to be used as input data of predictive models. These models must take into account mechanical and (electro)-chemical parameters. Significant advances are expected from numerical simulation in understanding material evolution and dissolution / repassivation mechanisms. Development of sophisticated models is also a key-aspect in the predictive understanding of tribocorrosion. The modelling of tribocorrosion is addressed further on in Chapter 5.

4.9 References

1. S. Mischler, A. Spiegel, M. Stemp and D. Landolt: *Wear*, 2001, **251**, 1295–1307.
2. S. Barril, S. Mischler and D. Landolt: *Tribol. Int.*, 2001, **34**, 599–608.

3. J. Takadoum: *Corros. Sci.*, 1996, **38**, 643–654.
4. S. W. Watson, F. J. Friedersdorf, B. W. Madsen and S. D. Cramer: *Wear*, 1995, **181–183**, 476–484.
5. F. Assi and H. Böhni: *Tribotest J.*, 1999, **6**, 17–28.
6. P. Jemmely, S. Mischler and D. Landolt: *Wear*, 2000, **237**, 63–76.
7. I. Garcia, D. Drees and J. P. Celis: *Wear*, 2001, **249**, 452–460.
8. X. X. Jiang, S. Z. Li, D. D. Tao and J. X. Yang: *Corrosion*, 1993, **49**, (10), 836–841.
9. R. Oltra: in 'Wear–Corrosion Interactions in Liquid Media', (eds. A. A. Sagües and E. I. Meletis), 3–17, 1991, Minerals, Metals and Materials Society.
10. D. Hamm, K. Ogle, C.-O. A. Olsson, S. Weber and D. Landolt: *Corrosion Science*, 2002, **44**, 1443–1456.
11. P. Schmutz and D. Landolt: *Electrochimica Acta*, 1999, **45**, 899–911.
12. P.-Q. Wu and J. P. Celis: 'Electrochemical noise measurements on stainless steel during corrosion–wear in sliding contacts', *Wear*, 2004, **256**, (5), 480–490.
13. P. Ponthiaux, F. Wenger and J. Galland: *J. Electrochem. Soc.*, 1995, **142**, (7), 2204–2210.
14. P. L. Andresen and F. P. Ford: *Corrosion Science*, 1996, **38**, 1011–1016.
15. Bom Soon Lee, Han Sub Chung, Ki-Tae Kim, F. P. Ford and P. L. Andersen: *Nuclear Engineering and Design*, 1999, **191**, 157–165.
16. M. Keddam M and W.R.Whitney: 'Award Lecture: Application of advanced electrochemical techniques and concepts to corrosion phenomena', *Corrosion*, 2006, **62**, 1056–1066.
17. M. Keddam, V. Vivier, D. Rose, P. Ponthiaux, F. Wenger and JP. Celis: 'Application of EIS in tribo-electrochemistry. State-of-the-art and perspectives', 8th International Symposium on Electrochemical Impedance Spectroscopy, Algarve, Portugal, June 2010.
18. E. Lemaire and M. Le Calvar: *Wear*, 2001, **249**, 338–344.
19. M. Stern and A. L. Geary: *J. Electrochem. Soc.*, 1957, **104**, 56.
20. N. Diomidis, J.-P. Celis, P. Ponthiaux and F. Wenger: *Wear*, 2010, **269**, 93–103.
21. M. Keddam, O. R. Mattos and H. Takenouti: *J. Electrochem. Soc.*, 1981, **128**, (2), 166–257.
22. F. Wenger and J. Galland: *Electrochim. Acta*, 1990, **35**, (10), 1573–1578.
23. Pei-Qiang Wu and J. P. Celis: *Wear*, March 2004, **256**, (Issue 5), 480–490.
24. Zhenlan Quan, Pei-Qiang Wu, Lin Tang and J.-P. Celis: *Applied Surface Science*, 30 November 2006, **253**, (Issue 3), 1194–1197.
25. A. Berradja, D. Déforge, R. P. Nogueira, P. Ponthiaux, F. Wenger and J. P. Celis. 'An electrochemical noise study of tribocorrosion processes of AISI 304 L in Cl⁻ and SO₄²⁻ media'. *Journal of Physics D, Applied Physics*, 2006, **39**, (n°15), 3184–3192.
26. D. Déforge, F. Huet, R. P. Nogueira, P. Ponthiaux and F. Wenger: 'Electrochemical noise analysis of tribocorrosion processes under steady-state sliding regime'. *Corrosion (NACE)*, 2006, **62**, (n°6), 514–521.
27. F. Wenger, P. Ponthiaux, L. Benea, J. Peybernès, A. Ambard: in 'Electrochemistry in Light Water Reactors', (eds. R.-W. Bosch and D. Féron, J.-P. Celis), 195–211, 2007, European Federation of Corrosion Publications, **49**, Woodhead Publishing and Maney Publishing, (ISSN 1354-5116).
28. N. Diomidis, J.-P. Celis, P. Ponthiaux and F. Wenger: *Wear*, 2010, **269**, 93–103.
29. P. Ponthiaux, F. Wenger, D. Drees and J.P. Celis: *Wear*, 2004, **256**, 459–468.
30. F. Galliano, E. Galvanetto, S. Mischler and D. Landolt: *Surface and Coatings Technology*, 2001, **145**, 121–131.
31. M. Azzi, M. Benkahoul, J. A. Szpunar, J. E. Klemberg-Sapieha and L. Martinu: *Wear*, 2009, **267**, 882–889.
32. M. T. Mathew, E. Ariza, L. A. Rocha, A. C. Fernandes and F. Vaz: *Tribology International*, 2008, **41**, 603–615.
33. J. Geringer, B. Normand, C. Alemany-Dumont and R. Diemiaszonek: *Tribology International*, 2010, **43**, 1991–1999.
34. M. M. Stack and K. Chi: *Wear*, 2003, **255**, 456–465.

35. M. M. Stack: *International Materials Reviews*, 2005, **50**, 1–17.
36. M. M. Stack, H. Jawan and M. T. Mathew: *Tribology International*, 2005, **38**, 848–856.
37. M. T. Mathew, P. Srinivasa Pai, R. Pourzal, A. Fischer and M. A. Wimmer: *Advances in Tribology*, doi:10.1155/2009/250986.
38. M. Grédiac: *Composites: Part A*, 2004, **35**, 751–761.
39. A. Kobayashi: 'Handbook on experimental mechanics, Weinheim: Society for Experimental Mechanics', 1999, VCH publishers, Inc.
40. J. C. Dainty: 'Laser speckle and related phenomenon', 1984, Berlin, Springer.
41. J. A. Leendertz: *J. Phys. E*, 1970, **3**, 214–218.
42. P. S. Theocaris: 'Moiré fringes in strain analysis', 1969, Elmsford, Pergamon press.
43. Post 94: 'High sensitivity moiré: experimental analysis for mechanics and materials', 1994, Berlin, Springer.
44. T. Kreis: 'Holographic interferometry: principles and methods', 1996, Berlin, Wiley-VCH.
45. M. A. Sutton, W. J. Wolters, W. H. Perters, W. F. Ranson and S. R. McNeill: *Image Vis. Comput.*, 1983, **1**, 133–139.
46. J. S. Sirkis: *Opt. Engng.*, 1990, **29**, 1485–1493.
47. Y. Surrel: in 'Interferometry '94: Photomechanics', (eds. Pryputniewicz RJ and Stupnicki J), 213–220, 1994, Berlin, Springer.
48. D. W. Templeton DW: 'Computerization of carrier fringe data acquisition, reduction and display'. *Exp Tech*, 1987, **11**, 26–30. Society for Experimental Mechanics.
49. J. Bulhak and Y. Surrel: in 'Interferometry '99: Techniques and Technologies', 20–23 September, Pultusk, Poland, vol. SPIE 3744, 1999, 506–15.
50. P. K. Rastogi: *Appl. Opt.*, 1998, **37**, 1292–1298.
51. J. J. Wasowski and L. M. Wasowski: 'Computer-based optical differentiation of fringe patterns'. *Exp Tech*, 1987, **11**, 16–18. Society for Experimental Mechanics.
52. D. Kempf, V. Vignal, G. Cailletaud, R. Oltra, J. C. Weeber and E. Finot: *Phil. Mag.*, 2007, **87**, 1379–1399.
53. V. Vignal, N. Mary, P. Ponthiaux and F. Wenger: *Wear*, 2006, **261**, 947–953.
54. H. Benabdallah and D. Olender: *Wear*, 2006, **261**, 1213–1224.
55. Y. N. Kok, R. Akid and P. Eh. Hovsepian: *Wear*, 2005, **259**, 1472–1481.
56. A. M. Simoes, A. C. Bastos, M. G. Ferreira, Y. Gonzalez-Garcıa, S. Gonzalez and R. M. Souto: *Corr. Sci.*, 2007, **49**, 726–739.
57. H. Lajain: *Werkst. Korros.*, 1972, **23**, 537.
58. H. Krawiec, V. Vignal and R. Akid: *Electrochim. Acta*, 2008, **53**, 5252–5259.
59. H. Krawiec, V. Vignal and R. Akid: *Surface and Interface Analysis*, 2008, **40**, 315–319.
60. H. Böhni, T. Suter and F. Assi: *Surface and Coatings Technology*, 2000, **130**, 80–86.
61. F. Assi and H. Böhni: *Wear*, 1999, **233–235**, 505–514.
62. P. Ponthiaux, F. Wenger, J. Galland, P. Kubecka and L. Hyspecka: Proceedings of the 14th International Corrosion Congress, CapeTown, South Africa, September 26–October 1, 1999. CD-ROM, Document Transformation Technologies, International Corrosion Council, Irene, South Africa. ISBN 0-620-23943-3.

5

Design of a tribocorrosion experiment on passivating surfaces: Modelling the coupling of tribology and corrosion

François Wenger

Ecole Centrale Paris, Dept. LGPM, F-92295 Châtenay-Malabry, France
francois.wenger@ecp.fr

Michel Keddam

Université Pierre et Marie Curie (Paris VI), Lab. L.I.S.E, CNRS UPR 15, F-75252 Paris, France
michel.keddam@upmc.fr

This chapter gives the pathway for designing a tribocorrosion experiment allowing separation of the mechanical, corrosive and synergetic contributions to the total loss of material. The general features of the tribocorrosion mechanism are presented emphasising the meaning and origin of the interplay between mechanical and reactive phenomena. A more detailed formulation of the problem is proposed in the framework of the active wear track. Based on this description, conditions are presented for performing a meaningful tribocorrosion experiment by combining tribological and electrochemical requirements. Emphasis is put on features specifically oriented towards a test based on the active wear track concept. A simplified tribocorrosion model is derived in terms of a repassivation function describing empirically a number of literature data in the field. The current responses to the various tribological regimes involved in the test are then derived and their relationships to the components of the main equation describing the test explained. Finally, the main limitations to the applicability of the test are discussed. The experiments to be implemented in this test are listed.

5.1   Various tests for various purposes

Three different approaches can be considered for the experimental study of tribocorrosion:

- A **fundamental study** aims at understanding the tribocorrosion mechanism and the kinetics of the resulting material degradation processes. Such research work is usually based on the use of a wide variety of laboratory tests and experimental techniques of data analysis. The tests are intended to characterise mechanical and physicochemical phenomena involved in tribocorrosion as well as the effects of tribocorrosion on the materials such as changes in mechanical and structural properties, electrochemical properties, corrosion and wear resistance, etc.
- An **experimental study** aims at evaluating and comparing the resistance of different materials to degradation induced by tribocorrosion. These studies are based on experimental techniques and test protocols specifically developed

to quantify the tribocorrosion behaviour of materials. Such studies can result in the development of standardised tests and are of great interest to industries or end-users of mechanical devices.

–   **Technological tests or bench tests** are designed to reproduce the mechanical loading and/or environment corresponding to actual operating conditions, or to mimic particular conditions intended to accelerate material degradation processes. These tests are widely used to predict precisely the behaviour of mechanical devices in actual conditions of service and to improve their reliability and durability. In this respect, they are very useful tools.

These three approaches provide complementary data in practice:

–   The fundamental research is driven by problems faced by real systems, by questions about the origin of their failures, the desire to improve their operations or to develop new mechanical systems.
–   The tests for evaluating the intrinsic properties of resistance to tribocorrosion are very useful insofar as they guide the engineers in selecting the materials for the design of new tribological systems or the improvement of existing ones. However, to be a valuable tool, they must be based on a sufficient knowledge of the characteristics of the tribocorrosion mechanism and associated kinetic laws established by fundamental research work.
–   The technological tests and bench tests are essential to the final stage of the development of tribological systems. They can be used under favourable conditions and at a minimal cost only if knowledge on the mechanism of tribocorrosion and on the behaviour of materials used, obtained by the two types of studies described above, is sufficient.

5.2   General features of the tribocorrosion mechanism

The tribocorrosion process involves a large number of phenomena resulting from the mechanical stressing of material surfaces and the action of the environment. Among them, the phenomena that have a major role in the behaviour of the material are considered hereafter and an attempt is made to understand the origin of the synergistic effect between corrosion and wear.

5.2.1   Main phenomena caused by friction and by the action of the aqueous medium

The action of an aqueous medium on the surface of a passivating material is expressed mainly by:

–   the formation of layers of adsorbed chemical species
–   the oxidation of the metal that leads to the formation of a passive surface film. Oxidation reactions can also induce a more or less transient dissolution of the material. Oxide layers or other reaction products can also be formed. Oxidation reactions are always involved in corrosion or passivation and are always associated with reduction reactions of, for example, hydrogen ions and/or oxygen dissolved in the medium.

The friction experienced by the surface of a material has an impact on its structural and mechanical state:

- The material undergoes a surface straining: in the contact zone, the material is subjected to compressive, tensile, and shear stresses which induce alterations in the stress state over a certain depth, with plastic deformation, cracking, abrasion, and microwelding resulting in adhesion and rupture.
- Straining by friction can also cause structural changes. For example, strain-induced martensitic transformation on AISI 304 steel.
- Friction also produces a partial or total destruction of the surface layers produced by the action of the aqueous medium on the contact areas, i.e. layers of adsorbed chemical species, thin films (passive film), oxide layers or layers of other solid products.

5.2.2 Origin of the synergy between corrosion and wear

The synergistic effect in tribocorrosion originates from the interactions between the effects of friction and those of the aqueous medium.

Influence of friction on the electrochemical behaviour of material surfaces

Friction can cause drastic changes in the electrochemical state of a surface and on the kinetics of reactions that take place:

- The destruction of surface layers by friction in contact areas induces a major perturbation of the kinetics of electrochemical reactions. In the case of passivating metals and alloys, the destruction of the passive film exposes bare metal to the surrounding oxidising medium. Oxidation reactions such as dissolution and passivation, are then activated. The overall rate of oxidation of bare metal areas can be multiplied by several orders of magnitude, thus accelerating the corrosive wear.
- The straining of rubbed areas and adjacent areas may alter the reactivity of a material. The importance of the resulting effect depends very much on the material and the intensity of straining. The presence of residual stresses changes the sensitivity of the material to corrosion.
- Friction results in a heterogeneity of the surface material, causing galvanic coupling phenomena at different scales, namely:
  - Galvanic coupling occurs between the sliding track where the passive film is destroyed, the strained material, and the surrounding area not strained by friction and where the passive film remains. Generally, the sliding track acts as the anode and the surrounding area as the cathode. This galvanic coupling has been studied experimentally and theoretically during [1] and after sliding [2].
  - Galvanic coupling can also occur at a more microscopic scale between different parts of the rubbed area having different structural and stress states. In addition, a partial restoration of the passive film may occur between two successive contacts with the counter body. At every time during friction, the proportion of restored passive film in a place on the sliding track is all the greater when the distance behind the sliding counterpart is high. This results in a galvanic coupling between sliding track areas located at different distances behind the pin. This galvanic coupling affecting the sliding track has been modelled in the case of a pin-on-disc experiment [1].

Influence of electrochemical phenomena on the tribological behaviour of the surface and on mechanical wear

Conversely, the electrochemical reactions involved in corrosion and passivation processes produce significant changes to the surfaces and therefore to the sliding conditions, and to the mechanical wear:

- The formation of surface layers results from electrochemical phenomena and has an important influence on friction and wear, since the composition, structure and mechanical properties of the surface are radically altered. The oxide layers often have a decisive influence not only on the coefficient of friction [3] but also on wear, due to a composition and mechanical properties that differ from those of the substrate. Depending on their hardness, brittleness, and their adhesion to the substrate, they may be more or less resistant to wear than the bare metal, and either constitute a protection against mechanical wear, or induce an acceleration thereof.
- Electrochemical reactions also contribute to the formation of the debris that promotes either the accommodation of speeds and reduction of wear, or acts as abrasive particles increasing mechanical wear.
- Despite their limited thickness of the order of interatomic distances, adsorbed layers can significantly alter the sliding conditions reducing the interaction forces between contacting surfaces.
- Corrosion in the rubbed areas may also contribute to the modification of the topography and roughness of the contacting surfaces, and the change in the shape of the contact, thus changing the sliding conditions, namely friction and wear.

5.3   General characteristics of the test: quantities to evaluate

The resistance of passivating materials to tribocorrosion involves the ability of that kind of materials to repair the passive film when mechanically damaged. The concept of the test (type of contact, nature of the counter body, and the relative motion of the two contacting bodies, etc.) as well as the test procedure (tribological conditions, electrochemical conditions, running of the test, etc.) and measurement techniques implemented *in situ* and *ex situ* have to be selected so as to obtain a sufficiently accurate assessment of the total wear, the mechanical and electrochemical wear components, and the quantification of the effect of synergy.

The total wear, W_t, is usually expressed as the sum of two components and, corresponding, respectively to the effects of mechanical, W^m, and electrochemical, W^c, phenomena:

$$W_t = W^m + W^c \qquad [5.1]$$

In the absence of sliding under the same environmental conditions, the material loss measured on the same area covered with passive film is generated only by electrochemical reactions, W_0^c.

Similarly, we can define the purely mechanical material loss that would occur if the area was rubbed under the same tribological conditions but in the absence of an aqueous environment: W_0^m.

The total wear under conditions of tribocorrosion can then be expressed as follows:

$$W_t = W_0^m + W_0^c + W^s \qquad [5.2]$$

This form of expression was proposed by several authors [4–6] where W^s is the term that reflects the synergy effect between mechanical and electrochemical phenomena detailed above. This term W^s is often expressed as the sum of two terms [7, 8]:

$$W^s = W^{mc} + W^{cm} \qquad [5.3]$$

W^{mc} is the modification of the mechanical material loss caused by the effect of electrochemical phenomena (formation of layers, roughness, etc.).
W^{cm} is the modification of the corrosive material loss caused by sliding effects (destruction of the surface layers, straining).

More precise information on the individual contributions of the bare metal and the passive film in the synergistic effect is necessary. This can be obtained by using an expression of the total wear W_t that is slightly different from the one given above and based on the concept of 'active wear track' [9]. This concept is particularly interesting to give a description of the state of the sliding track during a tribocorrosion test conducted on a pin-on-disc tribometer in which the pin is driven in a reciprocating linear motion or a continuous circular motion.

Each contact of the pin on the sliding track causes a partial or total destruction of the passive film in that contact area. On the bare metal area created by this destruction, oxidation occurs as soon as the contact with the pin comes to an end. This oxidation involves two competitive reactions that take place in parallel:

– a dissolution of the metal in the aqueous medium
– the re-growth of the passive film which probably occurs through nucleation, growth, and coalescence processes.

During the latency time, t_{lat}, defined as the time between two successive contacts at a given point in the sliding track, the passivation reaction tends to restore the passive film. The fraction of the sliding track surface covered by this re-grown passive film increases with increasing t_{lat}.

The component W^c corresponds to material loss induced by electrochemical phenomena. A simplified description of the electrochemical reactions leading to material loss can be given when the sliding track area is assumed to be divided into two distinct parts:

– an 'active' area of bare metal, A_{act}
– a passivated or repassivated area, A_{repass}, covered by a passive film which reformed since the last contact with the pin, and is possibly part of the surface film not destroyed on contact with the pin:

$$A_{tr} = A_{act} + A_{repass} \qquad [5.4]$$

with A_{tr} the total area of the sliding track.

The following dissolution processes and corresponding material losses can be considered, namely:

– the dissolution of the 'active' bare metal on A_{act}. The corresponding material loss is W_{act}^c
– the residual dissolution of passivated metal on A_{repass}. The corresponding loss of material by dissolution of this film is W_{repass}^c.

Then W^c is expressed as the sum of these two components.

It must be noted that such a 'binary' description of the electrochemical state of the sliding track is only a convenient tool for quantifying the different terms of W_t in Equation 5.1. A more realistic picture of the surface of the track should consider a continuous distribution of intermediate states between bare metal and a fully re-passivated surface state.

The component W^m corresponds to a material loss caused by mechanical phenomena. If the same simplified description of the state of the sliding track is considered, this term can be split in two components:

- W^m_{act} the material loss by mechanical degradation of bare metal on A_{act}
- W^m_{repass} the material loss by mechanical degradation on A_{repass}. This term includes the loss of passive film as well as a possible loss of underlying material.

It is interesting to assess from tribocorrosion tests, the importance of the synergistic effect caused by the interaction between the sliding and electrochemical phenomena. The term W^s that quantifies the synergy can be defined using the following reasoning:

- The aqueous medium causes the oxidation of the metal resulting in the dissolution of bare metal and the formation of passive film. In the absence of an aqueous medium, the material loss would be a 'pure' mechanical wear, W^m_{act}, of bare metal not covered by a passive film.
- Conversely, in an aqueous medium and in the absence of sliding, the metal is completely passivated and the material loss is a 'pure' electrochemical one, corresponding to the residual dissolution of the passive material, W^c_{repass}.

Under conditions of tribocorrosion, the synergy effect appears as the occurrence of a material loss corresponding to the destruction of the passive film by a mechanical action, W^m_{repass}, and a material loss due to the dissolution of active metal, W^c_{act}.

In pin-on-disc experiments, the synergy component W^s defined in Equation 5.2 can be defined as the loss of material resulting from the sum of two terms:

$$W^s = W^c_{act} + W^m_{repass} \qquad [5.5]$$

The expression of the total wear W_t is then given by [10, 11]:

$$W_t = W^c_{act} + W^c_{repass} + W^m_{act} + W^m_{repass} \qquad [5.6]$$

It must be noted that the different terms in this expression are defined according to the concept of active wear track, and differ significantly from the definitions given by other authors [4–8] for the terms in Equations 5.2 and 5.3. The components in Equation 5.6 correspond to material losses obtained during the tests on the fraction of bare, A_{act}, and repassivated, A_{repass}, areas. This weighting by the areas A_{act} and A_{repass} does not appear in the work cited above [4–6].

5.4   Different types of tests, design of a particular test

5.4.1   Choosing the tribometer

Many different tribological system concepts are used for sliding tests. Hereafter, the focus is on the tests that seem best suited for analysing and evaluating the resistance to tribocorrosion of passivating metals or alloys, and to highlight the role of the passive film.

The pin-on-disc test appears to be well suited. Indeed, if the surface of the disc is made of the passivating material to be tested, on sliding, the pin can partially or totally destroy the passive film in the contact area on the disc. After the contact with the pin, the passive film tends to re-build before the next contact event. At each contact event, the disc material undergoes a mechanical action applied on a partially depassivated surface and also an electrochemical action leading to the dissolution and repassivation of the bare metal. For that reason, pin-on-disc tribometers can be selected as a laboratory tribocorrosion test for passivating materials.

The most used tribometers of that type are either reciprocating tribometers or pin-on-disc tribometers.

Many tribocorrosion studies have been carried out with such tribometers, adapted to tribocorrosion tests in aqueous medium by installing the sample to be studied in a cell containing an electrolyte. Electrochemical measurements can be performed with both types of tribometer. However, to be implemented under conditions that allow the interpretation of results, some methods require static electrochemical conditions, at least before starting the measurement. Such conditions can be achieved on pin-on-disc tribometers operating at a constant rotation speed.

Among the electrochemical methods used in tribocorrosion studies performed with pin-on-disc testers, the following can be proposed:

- measurement of the open circuit potential and its evolution with time
- recording of current or potential transient responses induced by imposed variations of potential or current
- plot of potentiodynamic polarisation curves
- plot of electrochemical impedance spectra.

The last three methods are very powerful tools for analysing the mechanisms and kinetics of electrochemical reactions.

In a tribocorrosion test, various possible causes of non-stationary behaviour are:

- Non-stationary electrochemical reactions. This point is not specific to tribocorrosion tests.
- The pin movement which generates a perturbation (periodic or not) of the electrochemical state of the material on the sliding track [12].
- The gradually changing surface conditions (mechanical, structural, etc.) and the geometric characteristics of the sliding track (in the case of a non-conformal contact sphere–plane, for example, the width of the track increases over time).

Among these possible causes, a reciprocating movement of the pin actually introduces a periodic perturbation of the electrochemical conditions. With a reciprocating tribometer, the plot of potentiodynamic polarisation curves is skewed by periodic fluctuations of current induced by the motion of the pin. For similar reasons, impedance diagrams as well as transient responses to an imposed variation of potential or current, cannot be recorded in a satisfactory way. Only the variations of electrochemical parameters correlated with the movement of the pin can be analysed to study the behaviour of the surface and the contribution of electrochemical reactions to the material degradation.

On the other hand, a uniform rotation of the pin is not inconsistent with a stationary current or potential, although locally at any point in the sliding track, the electrochemical state of the surface is periodically perturbed by the contact event with the pin. If, when tested over a certain period, the area and the mechanical and structural states of the sliding track do not evolve significantly, a stationary open circuit potential, or current under potentiostatic conditions, can be observed. This has been verified experimentally many times [13] and can be justified theoretically, as will be shown in the last part of this chapter (cf. Section 5.5.4). Consequently, the use of a rotating pin-on-disc tribometer for tribocorrosion tests is preferable to that of a reciprocating tribometer, insofar as it allows us to consider the implementation of a greater number of electrochemical techniques for studying the role of corrosion and passivation in the degradation of a material, and to quantify their contribution. The concept of mounting a sample on a rotating pin-on-disc tribometer for tribocorrosion tests is illustrated in Figure 5.1.

5.4.2   Choosing the tribological test conditions

- **Antagonist materials**

  To characterise the resistance of the metallic material to tribocorrosion and its ability to undergo repassivation, an insulating and chemically inert pin is used. In this way, the chemical attack of the pin is avoided as well as the formation of a galvanic coupling between the pin and the material under investigation. This could profoundly alter the electrochemical behaviour of the latter. The material of the pin must also have specific mechanical properties: it may not deform too much under the applied load conditions, and should not undergo significant wear during the sliding test. This will limit the evolution of contact conditions and tribological conditions. The pin material should thus have a high modulus of elasticity and hardness in comparison to the tested material. Ceramics such as corundum (Al_2O_3), zirconia (ZrO_2), or silicon nitride (Si_3N_4), can be used as pin materials.

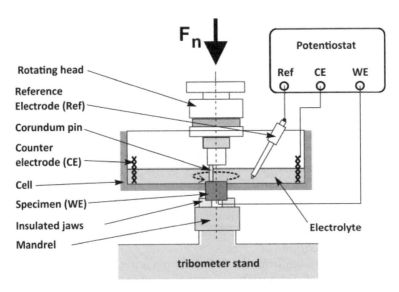

5.1 Concept of sample mounting on a rotating pin-on-disc tribometer for tribocorrosion tests

– *Contact conditions*

Preferably, a spherical pin or ball will be used. This configuration promotes contact conditions reproducible from experiment to experiment and stable during the movement of the pin along the sliding track, contrary to the plane-on-plane contact conditions. However, in the case of wear of the pin or ball, this kind of contact geometry causes an increase in the width of the sliding track over time, and thus an increase in the rubbed area. This variation causes a non-stationary electrochemical state, which should be considered for the implementation of electrochemical techniques and in the interpretation of the results.

The normal force applied between pin and disc surface must be sufficiently large to cause the material to be depassivated fully or partially at every contact event with the pin. However, it must not be too high so as to avoid excessive straining and mechanical deformation. The normal force can be selected such that the maximum stress near the surface and in the bulk of the material is below the compressive yield strength. For a sphere on plane system (ideally with smooth surfaces) and under the conditions that only elastic deformation is achieved, the contact area and stress distribution around the contact area can be calculated from Hertz theory as a function of the applied normal force, the radius of the sphere, and the elastic moduli of the materials in contact (see Chapter 3).

– *Sliding speed and contact frequency*

In a rotating pin-on-disc tribometer operating at a constant rotation rate, the sliding velocity, V_s, is related to the rotation period, t_{rot}, by the following equation:

$$V_s = \frac{2\pi R_{tr}}{t_{rot}} \qquad [5.7]$$

with R_{tr} the sliding track radius.

The mechanism and kinetics of material degradation may depend on the sliding speed, at least for certain ranges of speed. For example, an increase in sliding speed may promote an increase in the rate of material loss as was shown in tribo-corrosion tests carried out on a 304L steel in boric acid and lithium hydroxide solution (see Figure 5.2). On the other hand, the contact frequency directly affects the time t_{lat} during which the material can repassivate between successive contact events. It thus significantly influences the electrochemical behaviour of surfaces and the kinetics of the resulting tribocorrosion process.

To perform comparative tests between different materials, for example, it is preferable to choose identical sliding speeds and rotation periods. If any of these parameters must be changed, the influence of this change on the different components of the tribocorrosion process must be determined beforehand.

5.4.3   Principles of a possible procedure for performing a tribocorrosion test

The above-mentioned considerations can provide guidelines to clarify some general principles of a tribocorrosion test procedure that have to be developed:

– A rotating pin-on-disc tribometer will be used with an 'inert' pin (corundum, zirconia, etc.), with a spherical end. The studied material will constitute the disc (working surface) placed in a cell containing the aqueous medium and the electrodes required for electrochemical measurements.

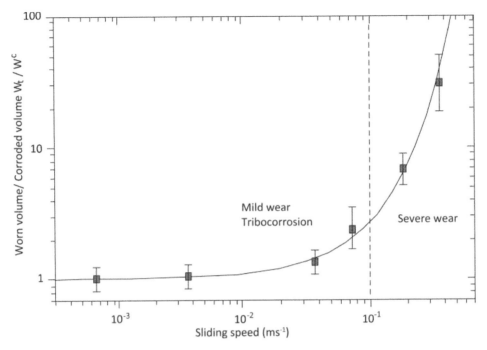

5.2 Influence of sliding speed on the ratio W_t/W^c. Tribocorrosion tests carried out with a pin-on-disc tribometer under continuous and intermittent sliding on AISI 304L stainless steel in a solution of boric acid and lithium hydroxide (1000 ppm B + 120 ppm Li; pH 8.2), from Ref. 27

–   A normal force will be applied to the disc resulting in a pressure sufficiently high to remove the passive film under sliding, but below the yield strength.

–   Sliding will be applied during a number of rotations sufficiently high to induce a significant and measurable wear.

–   In order to separate and quantify the contributions of the different components of the total wear, tests conducted under different sets of conditions will be necessary:

•   The components (W_{act}^c and W_{act}^m) can be first determined by an experiment in which the contribution of the two other components (W_{repass}^c and W_{repass}^m) are negligible. This can be achieved if sliding conditions prevent the restoration of the passive film on the sliding track. Such conditions are met when t_{lat} is sufficiently small to prevent the restoration of the film between two successive contacts.

•   An additional test is then needed to assess these last two components (W_{repass}^c and W_{repass}^m), with sliding conditions enabling a partial restoration of the passive film in the sliding track. It is possible to achieve such conditions by increasing t_{lat}. Two approaches are available:

first approach: The latency time t_{lat} can be modified by changing the rotation period t_{rot}. However, if the radius of the sliding track R_{tr} is kept constant, an increase in t_{rot} causes a proportional decrease in the sliding speed V_s which may change the mechanism of mechanical degradation as mentioned above. Another way to increase t_{lat} is to keep the sliding speed V_s constant and increase the radius of the track R_{tr}. However, for practical reasons, it is not realistic to consider increasing the radius of the track by a large factor.

second approach: The latency time is changed by performing *intermittent sliding* tests. During such an intermittent sliding test, the counter body slides for one rotation, and then stays immobile for a certain period of time to allow a part of the passive film to re-grow. Thus, an off-time, t_{off}, is imposed at the end of each rotation. As a result, the latency time between two subsequent contact events, t_{lat}, differs from the rotation period, t_{rot}:

$$t_{lat} = t_{rot} + t_{off} \qquad [5.8]$$

Sliding tests with $t_{off} = 0$ are referred to as 'continuous sliding tests'. Sliding tests with $t_{off} \neq 0$ are referred to as 'intermittent sliding tests'.

In these tests, t_{lat} is the period of a cycle (rotation step + stop step).

In intermittent sliding tests, sliding can be applied under the same conditions of sliding speed and track radius as those met in the continuous sliding tests.

This latter solution has been used in recent years to study the tribocorrosion of different passivating alloys, such as Stellite 6 [14] and 316 L steels [11].

Therefore, the different components of wear can be assessed using an experimental procedure including at least two tribocorrosion tests performed at two contact frequencies selected appropriately with respect to the time required to restore the passive film. The electrochemical components of the degradation process can be determined by electrochemical measurements carried out during the tests.

It must be noted that the implementation of an intermittent sliding test requires the use of a tribometer equipped with a stepping motor (with digital control, for example) designed to start and stop the rotation of the pin within a few rotation degrees, and capable of maintaining a constant speed equal to the set point value during the rest of the rotation. If such a tribometer is not available, the track radius has to be varied.

In the next section, a simplified model is presented to describe the electrochemical behaviour of a passivating material subjected to a tribocorrosion test in a rotating pin-on-disc tribometer. This model will serve as a basis for establishing a detailed test procedure (see Chapter 6) and to understand how to process experimental data for the evaluation of the different components of wear.

5.5   Simplified tribocorrosion model

5.5.1   Principles for the development of the model

In order to specify the experimental procedures for determining the components of tribocorrosion characterising the effect of synergy, and to provide a basis for the interpretation of the results of the test and the processing of the data, a tribocorrosion model must be developed. This model should describe the mechanism of tribocorrosion of passivating metals and alloys, including the phenomena at the origin of the synergy. This has to be done under the specific conditions corresponding to the type of tribological system which has been chosen (e.g. pin-on-disc tribometer).

As mentioned above, the synergy effect results from the effect of sliding on the electrochemical behaviour of the surface and the effect of the aqueous environment on the mechanical properties of the surface and the sliding conditions. The effect of sliding on the electrochemical behaviour of the surface is linked to:

- the destruction of the passive film, which reactivates the oxidation of bare metal, leading to dissolution and repassivation
- the straining of the surface which modifies its reactivity.

Some electrochemical phenomena can affect the mechanical properties of the surface and the contact conditions, such as:

- Modification of the mechanical resistance of the surface to sliding compared to that of the bare substrate. This modification results from the formation of a surface layer consisting of corrosion products or a passive film.
- Production of a third body in the contact with abrasive properties or on the other hand allowing the accommodation of forces and relative velocities.

However, the model should give a complete description of all mechanical and electrochemical aspects of the tribocorrosion mechanism including all phenomena at the origin of the synergy. Such a model would become too complex and probably of questionable generality to be used for interpreting test results given the large number of parameters characterising the system under study, and whose values are unknown *a priori*. Therefore, a simplified model of electrochemical behaviour of passivating materials subjected to sliding will be developed based mainly on a description of depassivation and repassivation phenomena which strongly influence the synergy. This model must give the relationships between the quantities to be determined (wear components), the parameters defining the test conditions (normal force, rotation period, t_{rot}, latency time, t_{lat}, etc.), and the measured quantities (total wear, electrochemical quantities, etc.).

Many electrochemical models have been proposed to account for the nucleation and growth of the passive film [15–18]. The 'High Field Conduction' type models [18] or derivatives offer varied and complex mechanisms explaining film growth and allowing deduction of the kinetic laws and the evolution of the latter with electrochemical conditions [19]. It seems impossible at this stage to propose a single comprehensive model for all passivating metals and alloys. Even if such a model could be developed, it is likely that its use for the interpretation of the tribocorrosion test that we are trying to develop would be too complex. As a result, a simplified model of depassivation–repassivation is proposed based on a semi-empirical approach and designed to provide an easy interpretation of test results and calculation of the degradation components.

5.5.2   First basic assumptions

As a starting point, the following assumptions are made for a pin-on-disc set-up as shown schematically in Figure 5.3:

- The working surface is represented by a disc of area A_0.
- The pin rubs on this disc, describing a circle of radius R_{tr}. The movement is performed at constant angular speed Ω.
- The sliding track area is A_{tr}:

$$A_{tr} = 2\pi R_{tr}\delta_{tr} \qquad [5.9]$$

where δ_{tr} is the width of the sliding track on which the material undergoes periodic contact with the pin.

- x is the distance between the pin and a given point on the wear track:

$$x = R_{tr}\theta \qquad [5.10]$$

with θ the angular distance between the pin and the point.

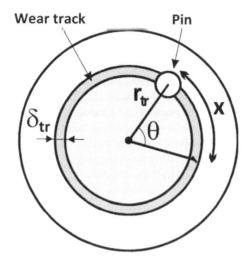

5.3 Pin-on-disc system

5.5.3   Depassivation–repassivation of the material

Some simplifying assumptions are made concerning the process of mechanical depassivation. It is assumed that, at every contact event with the pin, and in every place on the sliding track, the destruction of the passive film occurs identically throughout the contact area between pin and disc material as soon as the pin comes into contact with it. It is assumed that the bare metal in the depassivated surface undergoes oxidation through two reactions, namely:

- The dissolution of the components of the material.
- The growth of a new passive film.

Both reactions occur in parallel and in a competitive manner on the surface. This behaviour has been reported for different passivating materials [20, 21]. Depending on the tested materials and the environments in which they are immersed, the relative rates of these reactions vary. Moreover, the dissolution rate is high compared to the repassivation rate at short time intervals after depassivation. Then the dissolution rate decreases gradually as the passive film grows [20, 21]. In the long term, when sliding is stopped, the passive film will be restored along the whole sliding track.

   To calculate the volume of material electrochemically dissolved during sliding, it is important to know the kinetic law of the electrochemical reactions taking place on the track. The kinetic law that we need gives the evolution of the oxidation current flowing through the bare surface formed after contact with the pin, during repassivation. From this law, the amount of material dissolved is calculated from Faraday's law.

5.5.4   Oxidation current density distributions

- *Current density variation with time*
   For reasons explained above (Section 5.5.1), mechanistic passivation models are not suitable to derive simple and general analytical expressions of the evolution of oxidation currents and of the rates of electrochemical reactions (dissolution

and passivation) after mechanical depassivation. Therefore, faced with these difficulties, an empirical expression has to be used to represent the evolution of anodic current density versus time, $i(t)$, on the bare surface of a passivating material. This expression reflects the anodic current observed during the repassivation of passivating alloys in various cases: pitting corrosion or stress corrosion [22, 23], and tribocorrosion [24–26]. It was obtained with materials of different composition and seems to represent the kinetic law for a wide variety of cases.

If i is the current density on the surface of the wear track, the evolution of i versus time t after depassivation ($t = 0$) is given by:

$$i = i_0 \left(1 + \frac{t}{t_0} \right)^{-p} + i_{pass} \qquad [5.11]$$

i_0 is the current density at $t = 0$, and t_0 is a characteristic time constant. From the experimental results obtained with different materials by various authors, it seems that p varies from one material–environment system to another, but always with $p \leq 1$. The current density i_0 is a quantity that depends on the potential and the mechanism of the electrochemical reactions occurring on the material. i_{pass} is a constant current corresponding to the steady-state or quasi steady-state passivation current density.

Even though the total amount of electrical charge involved by its integration tends to infinity at long times, Equation 5.11 has a certain generality since it has been observed under various electrochemical conditions:

– An evolution of the current density following this kinetic law has been observed during the passivation of various metallic materials under imposed anodic polarisation. It has been observed on Stellite 6 [25] as well as on AISI 316 or 304 stainless steels [27], when the surface depassivated previously by a polarisation at a cathodic potential, is suddenly subjected to an anodic polarisation applied in the potential range where the alloy is passive. It was also observed in tribocorrosion tests performed under applied anodic polarisation where depassivation was caused by a mechanical action.

– It is important to note that the kinetic law of Equation 5.11 also explains the evolution of the electrochemical wear component (W^c) under conditions of intermittent sliding, where the open circuit potential is not constant and undergoes large variations over time (typically several hundred millivolts). During the sliding step, the depassivation of the surface causes a sharp drop in the open circuit potential. During the stop step, repassivation of the surface of the track occurs under conditions of galvanic coupling between the depassivated track and the area outside the track that remains in a passive state. The potential starts to increase towards the value corresponding to that of a uniformly passive surface.

The experimental kinetic law for electrochemical wear $W^c(t_{lat})$ can be explained by an evolution of the anodic current corresponding to Equation 5.11. This result was found for different passivating alloys [3, 24–26].

It should be noted however that the value of the exponent p found under these conditions of an intermittent test at the open circuit potential is lower than that found in tests of repassivation at anodic applied potential. This difference is probably because, unlike the previous case, during an intermittent test, the repassivation of the wear track occurs under conditions of variable potential. The smaller value of the exponent p corresponds to slower passivation kinetics.

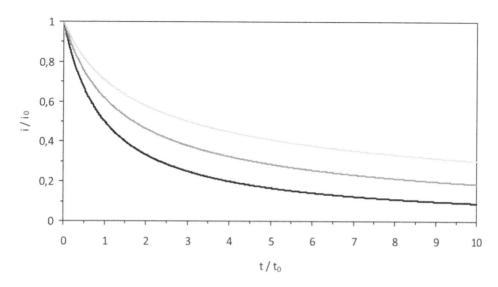

5.4 Repassivation current density transient in linear scales, according to Equation 5.11. Black line: $p = 1$; grey line: $p = 0.7$; light grey line: $p = 0.5$

For t sufficiently small compared to the time needed to achieve a steady passive state, the following expression can be considered:

$$i = i_0 \left(1 + \frac{t}{t_0} \right)^{-p}$$ [5.12]

This expression is valid for $i \gg i_{pass}$. In Figure 4.10 of Chapter 4, the repassivation transient obtained in a potential step experiment on Stellite 6 alloy in a 0.5 M H_2SO_4 solution shows that $i(t)$ decays as t^{-p} (with $p = 0.7$) between values of t from 0.1 s to a value larger than 100 s. Up to this last value of t, the steady current of passivation is still not reached and the Equation 5.12 is valid to describe the evolution of anodic current.

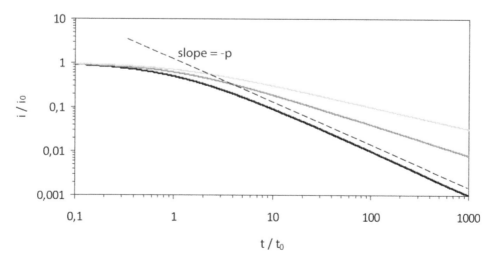

5.5 Repassivation current density transient in log–log scale, according to Equation 5.11. Black line: $p = 1$; grey line: $p = 0.7$; light grey line: $p = 0.5$

Then, assuming that the sliding track area is divided into active and repassivated areas (A_{act} and A_{repass}, respectively), it can be considered that Equation 5.12 corresponds to the evolution of the current density on the active area. The shape of this transient response is given in Figures 5.4 and 5.5 in linear and logarithmic scales, respectively.

– Current density distribution along the sliding track

If we consider the case of a pin-on-disc test performed at a constant rotation rate, it can be assumed that, at every point on the sliding track undergoing periodic contact, the current density follows the evolution given by Equation 5.12 between the time $t = 0$ corresponding to the time at which the contact with the pin is broken, and time $t = t_{rot}$, where t_{rot} is the period of rotation of the pin. The evolution of the current density can be expressed as a function of distance x from the point to the pin:

$$x = R_{tr}\Omega t = V_s t \qquad [5.13]$$

where t is the time elapsed since the last contact with the pin ($0 \leq t < t_{rot}$).

The expression for the spatial distribution of current density can be deduced at any time behind the pin arbitrarily located at $x = 0$:

$$i = i_0 \left(1 + \frac{x}{V_s t_0}\right)^{-p} \qquad [5.14]$$

This expression can also be written as:

$$i = i_0 \left(1 + \frac{t_{rot}}{t_0}\frac{x}{L}\right)^{-p} \qquad [5.15]$$

with L the length of the track ($L = 2\pi R_{tr} = V_s t_{rot}$).

It must be noted that as V_s is constant, the spatial distribution of i behind the pin is stationary. This distribution is represented in Figure 5.6.

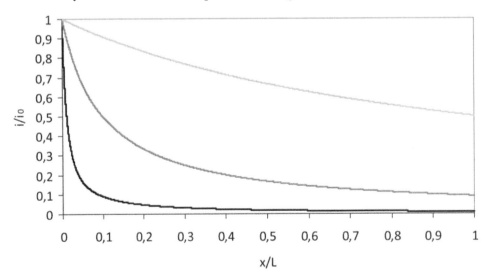

5.6 Current density distribution along the wear track during continuous sliding. x is the distance between the considered point of the track and the pin. Black line: $t_{rot}/t_0 = 100$; grey line: $t_{rot}/t_0 = 10$; light grey line: $t_{rot}/t_0 = 1$

5.5.5   Current under continuous sliding

The total current I obtained from the track is stationary under such sliding test conditions, and it can be easily calculated from:

$$I = \int_0^{A_{tr}} i\, dA = \delta_{tr} \int_0^L i_0 \left(1 + \frac{t_{rot}}{t_0}\frac{x}{L}\right)^{-p} dx \qquad [5.16]$$

which, after integration and if $p \neq 1$, becomes:

$$I = \frac{\delta_{tr} L\, t_0\, i_0}{(1-p)\, t_{rot}} \left[\left(1 + \frac{t_{rot}}{t_0}\right)^{(1-p)} - 1\right] \qquad [5.17]$$

As $A_{tr} = L\delta_{tr}$, and with $p \neq 1$, Equation 5.17 becomes:

$$I = \frac{A_{tr}\, i_0\, t_0}{(1-p)\, t_{rot}} \left[\left(1 + \frac{t_{rot}}{t_0}\right)^{(1-p)} - 1\right] \qquad [5.18]$$

If $p = 1$, Equation 5.18 takes the following form:

$$I = \frac{A_{tr}\, i_0\, t_0}{t_{rot}} \ln\left(1 + \frac{t_{rot}}{t_0}\right) \qquad [5.19]$$

The variation of I with t_{rot} is represented in Figure 5.7

The material under continuous sliding conditions at constant rotation speed is in an electrochemical steady state, since, if the potential is stationary, then according to the equations above, the current is also constant.

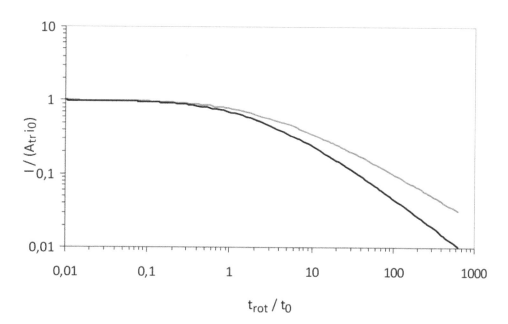

5.7 Steady-state conditions under continuous sliding. Variation of the ratio $I/(A_{tr}i_0)$ with t_{rot}/t_0. I is the steady-state current under continuous sliding. Black line: $p = 1$; grey line: $p = 0.7$

When continuous sliding is performed at constant rotation speed on passivating materials, the following experimental observations are made:

– Under conditions of open circuit potential, the measured potential often shows a long-term evolution not related to perturbations of the electrochemical state induced by the movement of the pin, but due to a slow broadening of the sliding track over testing time. This increase in the width of the sliding track slowly changes the conditions of galvanic coupling between active and passive surfaces. Fluctuations can also be observed related to the frequency of rotation of the pin, but that could be attributed to changes in normal force due to irregularities in the disc surface or a lack of parallelism between the plane of rotation of the end of the pin and the disc surface.
– Under conditions of applied potential, the evolution or fluctuations of the current that may be observed can be attributed to the same causes.

In fact, it can be concluded that:

– At a microscopic local scale, steady-state conditions are not met in pin-on-disc tests. Every point on the sliding track undergoes periodic fluctuations of potential and current associated with successive depassivation events resulting from the successive contacts with the pin.
– At a macroscopic scale however, at an applied potential or even under open circuit potential conditions, the rotation of the pin at constant rate tends to generate electrochemical conditions which correspond to an overall macroscopic steady state (total current and open circuit potential independent of time), despite the fact that this steady state can be disturbed by possible fluctuations of the normal force, and a long-term broadening of the sliding track.

5.5.6   Current under intermittent sliding

In order to avoid unwieldy demonstrations, only the expressions for $p \neq 1$ are considered in the next calculations. Calculations for $p = 1$, from Equation 5.12 would lead to similar results and the same conclusions.

The evolution of the current during the intermittent test is calculated separately for the sliding step and the stop step. As mentioned above, the validity of Equation 5.12 can be extended to the case of intermittent test conditions during which the potential evolves during the stop steps.

– Sliding step

The starting points are:

– At time $t = 0$, the pin is located at $x = 0$ (origin on the track).
– At time t ($t < t_{rot}$), it reaches the distance $\lambda = V_s t$ ($\lambda < L$).
– Behind the pin, for $0 \leq x \leq \lambda$, the distribution of current density takes the following form:

$$i_1 = i_0 \left(1 + \frac{t}{t_0} - \frac{x}{V_s t_0} \right)^{-p} \qquad [5.20]$$

or:

$$i_1 = i_0 \left(1 + \frac{t}{t_0} - \frac{t_{rot}}{t_0}\frac{x}{L}\right)^{-p}$$ [5.21]

In front of the pin, for $\lambda < x < L$, the distribution of i is that arising from the contact event with the pin during the previous rotation (except for the first rotation). The time elapsed from this last contact is t_{lat}. This current density distribution can be written as:

$$i_2 = i_0 \left(1 + \frac{t + t_{lat}}{t_0} - \frac{t_{rot}}{t_0}\frac{x}{L}\right)^{-p}$$ [5.22]

The distribution $i = i_1 + i_2$ at time t is represented in Figure 5.8.

The expression of the total current $I(t)$ during the sliding step is obtained by integrating the distribution of current density across the surface of the track:

$$I(t) = \delta_{tr} \left[i_0 \int_0^\lambda \left(1 + \frac{t}{t_0} - \frac{t_{rot}}{t_0}\frac{x}{L}\right)^{-p} dx + i_0 \int_\lambda^L \left(1 + \frac{t + t_{lat}}{t_0} - \frac{t_{rot}}{t_0}\frac{x}{L}\right)^{-p} dx\right]$$ [5.23]

as $A_{tr} = L\,\delta_{tr}$, the following expression is obtained:

$$I(t) = \frac{A_{tr}\,i_0\,t_0}{(1-p)\,t_{rot}} \left[\left(1 + \frac{t}{t_0}\right)^{(1-p)} - 1 + \left(1 + \frac{t_{lat}}{t_0}\right)^{(1-p)} - \left(1 + \frac{t + t_{lat} - t_{rot}}{t_0}\right)^{(1-p)}\right]$$ [5.24]

The evolution of I during the sliding step is represented for two values of t_{lat} in Figure 5.9.

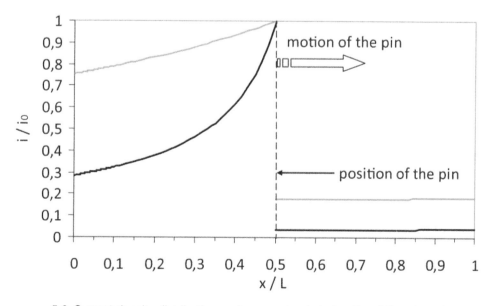

5.8 Current density distribution on the wear track during the sliding step of an intermittent tribocorrosion test with $t_{lat}/t_{rot} = 10$ and $p = 0.7$. Black line: $t_{rot}/t_0 = 10$; Grey line: $t_{rot}/t_0 = 1$

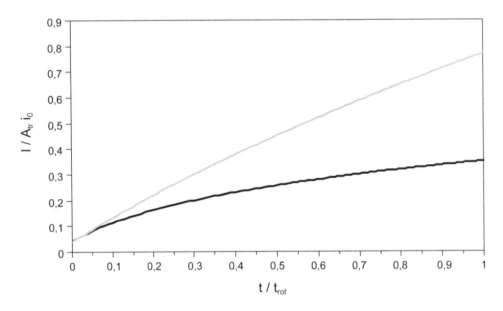

5.9 Evolution of the current flowing from the wear track during the sliding step of an intermittent test. Black line: $t_{rot}/t_0 = 10$, $t_{lat}/t_{rot} = 10$; grey line: $t_{rot}/t_0 = 1$, $t_{lat}/t_{rot} = 100$

The initial value of $I(t = 0)$ is all the smaller as t_{lat} is high. The value of I at the end of the rotation ($t = t_{rot}$) reaches the steady-state value of I obtained in a continuous sliding test regardless of the value of t_{lat}.

Stop step

During the stop step, the distribution of current density i across the sliding track at times $t > t_{rot}$ is no longer disturbed by the movement of the pin and evolves over time according to Equation 5.20:

$$i = i_0 \left(1 + \frac{t}{t_0} - \frac{x}{V_s t_0} \right)^{-p}$$ [5.25]

with $t_{rot} \leq t \leq t_{lat}$.

Equation 5.25 can also be written as follows:

$$i = i_0 \left(1 + \frac{t}{t_0} - \frac{t_{rot}}{t_0} \frac{x}{L} \right)^{-p}$$ [5.26]

This current density distribution is represented for different values of t in Figure 5.10.

The expression of the current $I(t)$ can be calculated by integrating this distribution over the surface of the sliding track:

$$I = \delta_{tr}\, i_0 \int_0^L \left(1 - \frac{x}{V_s t_0} + \frac{t}{t_0} \right)^{-p} dx$$

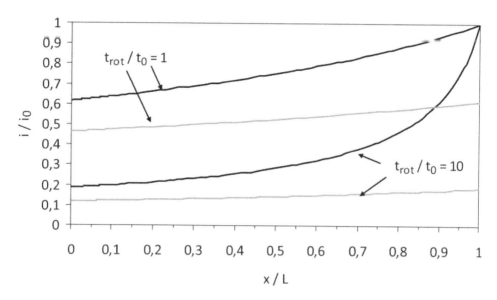

5.10 Current density distribution on the wear track during the stop step of an intermittent tribocorrosion test. $p = 0.7$. Black lines: distributions at $t = t_{rot}$ (end of rotation); grey lines: $t = 2t_{rot}$

$$I(t) = \frac{A_{tr}\, i_0\, t_0}{(1-p)\, t_{rot}} \left[\left(1+\frac{t}{t_0}\right)^{(1-p)} - \left(1+\frac{t}{t_0}-\frac{t_{rot}}{t_0}\right)^{(1-p)} \right]$$ [5.27]

This variation of $I(t)$ during the stop step is represented in Figure 5.11. It must be noted that, if $t \gg t_0$ and t_{rot}, the current decreases as t^{-p}.

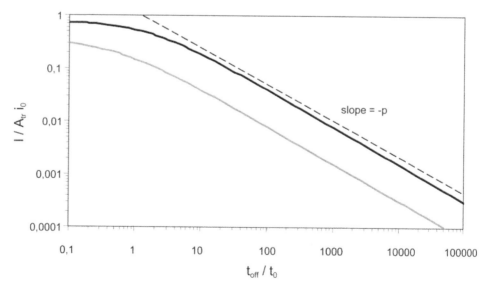

5.11 Evolution of the average current density during the stop step of an intermittent test. Black line: $t_{rot}/t_0 = 1$; grey line: $t_{rot}/t_0 = 10$

During continuous sliding tests performed at imposed anodic polarisation, the variation of I after stopping the tribometer can be recorded. The variation of $I(t)$ as t^{-p} can be verified and the value of p can be determined. However, the value of p determined in this way is only valid for repassivation at constant applied anodic potential. As mentioned above, i should also vary as t^{-p} under open circuit potential conditions, since Equation 5.12 is valid as mentioned above, but the value of p is different.

These expressions can be used to calculate the electrochemical material loss on the depassivated surface.

5.5.7   Charge consumed by oxidation of the material during tribocorrosion tests

From Equation 5.19 for continuous sliding, and Equations 5.24 and 5.27 for intermittent sliding, the electrochemical material loss in the sliding track can be calculated. The first step is to calculate the charge consumed by the anodic oxidation current on the track. Then the amount of metal oxidised can be derived by applying Faraday's law. The last step is to assess the proportion of dissolved metal and of metal oxidised to form the passive film.

Under conditions of continuous sliding

The current being steady during sliding, the corresponding charge Q per cycle (1 rotation) is given by:

$$Q = I\, t_{rot} \qquad [5.28]$$

with $p \neq 1$:

$$Q = \frac{A_{tr}\, i_0}{(1-p)}\, t_0 \left[\left(1 + \frac{L}{V_s t_0}\right)^{(1-p)} - 1 \right] \qquad [5.29]$$

or:

$$Q = \frac{A_{tr}\, i_0}{(1-p)}\, t_0 \left[\left(1 + \frac{t_{rot}}{t_0}\right)^{(1-p)} - 1 \right] \qquad [5.30]$$

Under conditions of intermittent sliding

The charges consumed during sliding (t_{rot}) and during the stop step (t_{off}) can be calculated from Equations 5.24 and 5.27.

Charge consumed during the sliding step

The charge consumed during sliding can be calculated by integrating Equation 5.24:

$$Q = \int_0^{t_{rot}} I(t)\, dt$$

The following expression is obtained:

$$Q(t_{rot}) = \frac{A_{tr}\, i_0\, t_0}{(1-p)}\left[\frac{t_0}{(2-p)t_{rot}}\left[\left(1+\frac{t_{rot}}{t_0}\right)^{(2-p)}-\left(1+\frac{t_{lat}}{t_0}\right)^{(2-p)}+1\right.\right.$$
$$\left.\left.+\left(1+\frac{t_{lat}}{t_0}-\frac{t_{rot}}{t_0}\right)^{(2-p)}\right]+\left(1+\frac{t_{lat}}{t_0}\right)^{(1-p)}-1\right]$$

[5.31]

It is important to note that the charge varies as $t_{lat}^{(1-p)}$ when $t_{lat} \gg t_0$ and t_{rot}.

Charge consumed during the stop step ($t_{rot} < t < t_{lat}$)

The charge consumed during sliding can be calculated by integrating Equation 5.27:

$$Q = \int_{t_{rot}}^{t_{lat}} I(t)\, dt$$

The following expression is obtained:

$$Q(t_{off}) = \frac{A_{tr}\, i_0\, t_0^2}{(1-p)(2-p)t_{rot}}\left[\left(1+\frac{t_{lat}}{t_0}\right)^{(2-p)}-\left(1-\frac{t_{rot}}{t_0}+\frac{t_{lat}}{t_0}\right)^{(2-p)}+1-\left(1+\frac{t_{rot}}{t_0}\right)^{(2-p)}\right]$$

[5.32]

For a given value of t_{rot}, if $t_{lat} \gg t_0$ and t_{rot}, $Q(t_{off})$ varies as $t_{lat}^{(1-p)}$.
During an intermittent test, the total charge per cycle is given by:

$$Q(t_{lat}) = Q(t_{rot}) + Q(t_{off})$$

[5.33]

It must be noted that $Q(t_{lat})$ is simplified as follows:

$$Q(t_{lat}) = \frac{A_{tr}\, i_0}{(1-p)}\, t_0\left[\left(1+\frac{t_{lat}}{t_0}\right)^{(1-p)}-1\right]$$

[5.34]

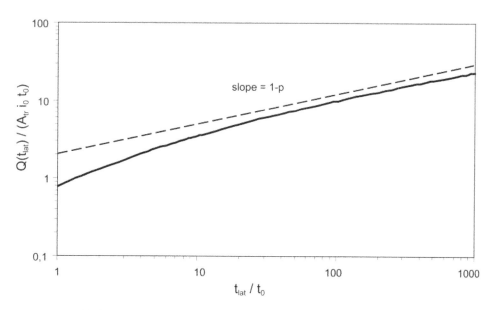

5.12 Variation of the total charge per cycle $Q(t_{lat})$ consumed during an intermittent test as a function of t_{lat}/t_{rot}

The total charge per cycle varies as $t_{\text{lat}}^{(1-p)}$ if $t_{\text{lat}} \gg t_0$. Therefore, the total charge per cycle $Q(t_{\text{lat}})$ and the electrochemical mass loss which is proportional also vary with t_{lat} in this way.

The variation of $Q(t_{\text{lat}})$ is represented in Figure 5.12.

5.5.8   Calculation of wear components

Continuous sliding

If under continuous sliding, the rotation period t_{rot} is sufficiently small, the current density everywhere on the wear track, calculated from Equation 5.15, is several orders of magnitude higher than the passivation current density. The whole wear track is considered as active ($A_{\text{act}} = A_{\text{tr}}$) and i is considered as a dissolution current density i_{act}.

Under these conditions, Equation 5.5 for the total wear in the wear track can be written as:

$$W_t = W_{act}^c + W_{act}^m \qquad [5.35]$$

The electrochemical component W_{act}^c (expressed as volume lost per cycle, for example) can be calculated by applying Faraday's law:

$$W_{act}^c = C_F \, Q_{act} \qquad [5.36]$$

where Q_{act} is the charge consumed during one rotation given by Equation 5.28 or 5.30. C_F is given by:

$$C_F = \frac{1}{F \, \rho} \left(\sum_j \left(\frac{x_j \, n_j}{M_j} \right) \right)^{-1} \qquad [5.37]$$

with F the Faraday's constant (96 500 C), ρ the density of the material, x_j and M_j, respectively the relative contents in weight and the atomic masses of the different metals j constituting the alloy, and the n_j the valency of the corresponding cations released into solution.

Then the mechanical wear component W_{act}^m can be calculated as the difference:

$$W_{act}^m = W_t - W_{act}^c \qquad [5.38]$$

The value of W_t is determined from profilometric or microtopographic measurements performed on the disc after the test.

Intermittent sliding

Under intermittent sliding, if the latency time is sufficiently long, it is possible to assume that the passive film is restored on a part of the wear track. This assumption is consistent with the results of measurements of the coefficient of friction carried out during tribocorrosion tests under continuous and intermittent sliding on passivating alloys. It was shown on passivating complex metal alloys [3], and AISI 316L stainless steel [10] that an increase in the coefficient of friction μ is observed when the latency time increases and seems to be linked to the increasing sliding contact area covered by re-grown oxide.

Thus, it is assumed that, during sliding and stop steps, the sliding track consists of two distinct zones, as expressed by Equation 5.4, namely:

- the active area, A_{act}, on which the material may be either bare, or covered by a layer which has not regained the mechanical and chemical properties of the original passive film.
- the remaining wear track area, A_{repass}, is covered by a surface film that is in the same state as before sliding. This film is either not removed by the counter body during sliding, or it had time to be restored to its initial state.

The advantage of using an intermittent test is to be able to change the time of repassivation of the surface of the wear track while retaining the loading conditions of the tribocorrosion test carried out under continuous sliding. It is thus possible to use some results of the test under continuous sliding to process the results of the intermittent sliding test. In particular, it can be assumed that, under intermittent sliding ($t_{lat} > t_{rot}$), the mechanical degradation on the active part of the track A_{act} is the same as under continuous sliding ($t_{lat} = t_{rot}$). Therefore, the mechanical degradation component under intermittent sliding ($W_{act}^m(t_{lat} > t_{rot})$) can be calculated from the mechanical degradation under continuous sliding ($W_{act}^m(t_{lat} = t_{rot})$) as follows:

$$W_{act}^m(t_{lat} > t_{rot}) = W_{act}^m(t_{lat} = t_{rot}) \frac{A_{act}(t_{lat} > t_{rot})}{A_{tr}}$$ [5.39]

remembering that $A_{act}(t_{lat} = t_{rot}) = A_{tr}$.

The corrosive component of tribocorrosion per cycle during the intermittent test cannot be measured. However, it can be assessed as follows:

As a first approximation, we can consider that the corrosive wear component W_{act}^c varies in proportion to $t_{lat}^{(1-p)}$, as shown by Equation 5.34. The validity of this approximation becomes better if the ratio t_{lat}/t_0 increases. The corrosive wear component under intermittent sliding, $W_{act}^c(t_{lat} > t_{rot})$, is then given by:

$$W_{act}^c(t_{lat} > t_{rot}) = W_{act}^c(t_{lat} = t_{rot}) \left(\frac{t_{lat}}{t_{rot}}\right)^{(1-p)} \frac{A_{act}(t_{lat} > t_{rot})}{A_{tr}}$$ [5.40]

It is worth noting that this prediction of the model is in agreement with the experimental variation of the corrosive wear with t_{lat} found with Stellite 6 [14, 25], and stainless steels subjected to intermittent sliding tests or intermittent sliding operating conditions [11].

The major issue at this stage is that the ratio $A_{act}(t_{lat} > t_{rot})/A_{tr}$ is unknown. There is no effective method to determine experimentally the fraction A_{repass}/A_{tr} of repassivated track area, i.e. covered by a passive film with almost the same protective properties as the initial film.

The value of the ratio A_{repass}/A_{tr} increases with the latency time t_{lat}. The calculation of this ratio by the model would lead unfortunately to a poor estimation because of the lack of knowledge of the values of i_0 and t_0 in particular. A first approach to this problem was proposed based on a crude assumption concerning the variation of A_{repass}/A_{tr} with time (see Equations 5.3, 5.11, and 5.12). The following relationship was taken:

$$\frac{A_{repass}}{A_{tr}} = \frac{t_{lat}}{t_{reac}}$$ [5.41]

where t_{reac} is the time consumed after initial immersion of the sample to reach quasi steady-state passivation (evolution of the open circuit potential lower than 60 mV h^{-1}). Until now, in previous works, intermittent tests were carried out with values of t_{lat} corresponding to very small values of A_{repass}/A_{tr} equal to 10^{-2} and 10^{-3}. However, for latency times of this magnitude, it was found that the coefficient of friction was significantly higher during sliding steps of intermittent tests than in continuous sliding tests. We also observed that the coefficient of friction increased with the latency time. As mentioned above, the increase in the coefficient of friction could be an indication of recovery of a portion of the surface by a passive film. In the test, the rotation period t_{rot} was chosen small enough compared to t_{reac} so as to neglect the fraction of repassivated wear track area ($A_{repass}/A_{tr} = 10^{-4}$) under conditions of continuous sliding ($A_{act} = A_{tr}$).

In Equation 5.6, the last two terms related to corrosive and mechanical wear, W^c_{repass} and W^m_{repass}, of the repassivated area of the track (A_{repass}) remain to be evaluated.

An approximate value of W^c_{repass} can be calculated if we consider that, on the repassivated fraction of the wear track area, the oxidation current density value has the same order of magnitude as the value i_{pass} deduced from polarisation resistance measurements at the beginning of the test with the sample in a uniform passive state. Because of the small value of this current density compared to the current density on the active part of the track, and because of the small values of A_{repass}/A_{tr}, the term W^c_{repass} is generally small enough to be neglected.

$$W^c_{repass} = C_F \, A_{repass} \int_0^{t_{lat}} i_{pass} \, dt \qquad [5.42]$$

C_F is calculated from Equation 5.37 considering the metallic element(s) which are components of the passive film.

The last term to be determined is W^m_{repass} which is the mechanical wear of the restored passive film. This term is calculated through Equation 5.6 as the difference between the total wear measured W_t and the sum of the three terms $W^c_{act} + W^m_{act} + W^c_{repass}$.

A decomposition of total wear W_t as that given by Equation 5.6 is used to specify the relative weight of the terms of corrosive wear and mechanical wear in the total wear for different physicochemical and tribological conditions. More particularly, for passivating materials, the intermittent sliding test associated with a simple electrochemical model and assumptions proposed for the repassivation of the wear track provide data for an assessment of the role of the passive film and of its restoration in the kinetics of tribocorrosion and wear.

Some variables may be defined to characterise more quantitatively the influence of various phenomena involved in the process of tribocorrosion on the wear. For example, the ratio K_c given by:

$$K_c = \frac{W^c_{act} + W^c_{repass}}{W^m_{act} + W^m_{repass}} \qquad [5.43]$$

can be used to evaluate the relative influence of corrosive and mechanical wear (Equation 5.11).

The ratio K_m was also defined (Equation 5.11) as:

$$K_m = \frac{W^m_{act}}{W^m_{repass}} \frac{A_{repass}}{A_{act}} \qquad [5.44]$$

which is the ratio of 'specific' mechanical wear components (W_{act}^m/A_{act} and W_{repass}^m/A_{repass}).

The value of this ratio is not useful in itself since it depends strongly on the value of A_{repass}/A_{tr} which is based on an arbitrary assumption concerning the rate of repassivation of the surface of the track with latency time. However, the change in the K_m value with t_{lat}, experimentally determined from several intermittent tests performed with different latency times, can highlight the influence of the formation of the passive film on the resistance of the surface to mechanical wear conditions of intermittent sliding, and can show if the film is more or less resistant than the metallic substrate to mechanical wear.

5.5.9   Limits of the model and further developments

The extreme complexity of the mechanism of tribocorrosion makes it difficult to give a description that takes into account all of the physicochemical and mechanical phenomena involved and that can lead to understanding and predicting the mechanism and kinetics of wear for a wide variety of passivating materials (metals and alloys) under various conditions of the environment and tribological operation.

Therefore, we tried to propose a simple testing procedure to characterise the resistance of different passivating materials. To design the test procedure and interpret the results, we proposed a very simple model which describes the evolution of the current during repassivation and the current distribution on the sliding track. This model is based on empirical observations of the variation of the repassivation current performed on different materials, which allows us to apply it to a great number of material/electrolyte systems. However, this requirement of generality and ease of use, does not allow us to base the model on detailed mechanisms of electrochemical reactions and growth of the passive film.

– Such a simplistic description of the electrochemical behaviour of the material does not allow us to describe in a realistic manner the kinetics of repassivation of the sliding track. According to the definition of a repassivated surface given above, the model suggests that the criterion for considering that the film was restored at a point on the track is that the current density has decreased below a threshold value i_p of the order of magnitude of the current density of passivation i_{pass} measured before applying sliding. In this case, the model shows that during a continuous or intermittent sliding test, the distance x_p between the pin and the point of the wear track where $i = i_p$ can be calculated from the expression of the current density distribution behind the pin ($x = 0$ taken as the distance behind the pin) at time t:

$$i = i_0 \left(1 + \frac{t}{t_0} + \frac{t_{rot}}{t_0} \frac{x}{L} \right)^{-p} \tag{5.45}$$

For $i = i_p$, we obtain:

$$x_p = V_s\, t_0 \left[\left(\frac{i_0}{i_p} \right)^{\frac{1}{p}} - 1 \right] - V_s\, t \tag{5.46}$$

The boundary between active and passive areas at a distance x_p behind the pin appears on the wear track at $x = L$ after an 'incubation' time t_p easy to calculate from Equation 5.46:

$$t_p = t_0 \left[\left(\frac{i_0}{i_p} \right)^{\frac{1}{p}} - 1 \right] - t_{rot} \qquad [5.47]$$

As shown by Equation 5.46, this boundary is moving linearly towards the pin as a function of time at a speed equal to the sliding speed V_s. As a result, the time required for total repassivation of the wear track is t_{rot} ($=L/V_s$) after the 'incubation' time t_p, when the entire wear track is active.

Such a prediction seems to be far from the experimental evidence: the difference between the values of i_p and i_0, deduced from experimental data, is generally of several orders of magnitude, giving values of t_p a hundred or a thousand times higher than those suggested by the results of the experimental study.

Three causes may give rise to this inconsistency:
- The concepts of active surface and repassivated surface should be revised, not based on a criterion such as the threshold value of i but, for example, on significant differences between electrochemical and mechanical properties and behaviour under sliding tests.
- The model proposed above is unidimensional. It only describes the distribution of current density in the longitudinal direction without considering possible current and potential distribution in the direction transverse to the track. In fact, because of the galvanic coupling effect between the track and adjacent passive areas, an important potential gradient exists on the edges of the track and film growth can take place not only in the longitudinal direction but also in the transverse one.
- The model could not be based on a passivation mechanism which would explain how the nucleation and growth of the passive film occur on the surface of the track, taking into account the electrochemical conditions of galvanic coupling, current–potential relationships, ohmic drop effects, and geometrical features of the system consisting of the track and the surrounding surface. Such a model should explain the variations of current with time (of type t^{-p}) observed in tribocorrosion studies as well as in other cases of localised corrosion, under conditions of constant applied potential or galvanic coupling. Until now, the mechanisms proposed for the growth of the passive film (high field models) do not predict such a variation, except for very specific values of parameters that cannot be justified in general.

Therefore, a rough assumption concerning the variation of repassivated area with time had to be proposed, so as to make the predicted behaviour more consistent with experimental evidence. To obtain a more detailed description of the electrochemical behaviour of the disc under continuous or intermittent sliding, several issues concerning repassivation should be addressed in further studies.

5.6   Measurements and techniques to be implemented in the test

For quantifying the different components of wear, a test procedure comprising experiments performed under continuous sliding and experiments under intermittent sliding (or continuous sliding at 'low' rotation rate) can be considered.

During these experiments, different *in-situ* techniques can be used:
Open circuit potential measurement gives information on the electrochemical state of the metal:

- At the beginning of the experiments, before applying sliding, the evolution of the open circuit potential indicates whether the metal is passivating and when a steady state has been reached.
- In a continuous or intermittent sliding experiment, when sliding is applied, the drop of the open circuit potential shows the depassivation of the sliding track. Under continuous sliding, the open circuit potential variation indicates whether a quasi steady state has been reached. Under intermittent sliding, following the potential fluctuations is a way to verify that the repassivation occurs in the same way during the stop phases throughout the test, and that depassivation and corrosion do not spread on the sample out of the sliding track.

As explained in Section 5.5.8, a tribocorrosion experiment under continuous sliding at a rotation rate sufficiently high to prevent repassivation of the sliding track can provide data for calculating the wear components W_{act}^c and W_{act}^m. During this experiment, because of the steady electrochemical state of the disc, EIS measurements can be implemented for measuring the polarisation resistance R_p. The anodic current I_{act} on the sliding track can be deduced from R_p as explained in the preceding chapter, and the corrosive wear component W_{act}^c can be calculated from Equation 5.28.

Then, the mechanical wear component W_{act}^m is calculated from Equation 5.38 from the value of total wear W_t measured after the end of the experiment.

For evaluating the wear components related to the passive film, W_{repass}^c and W_{repass}^m, intermittent sliding tests or continuous sliding tests at lower rotation rates can be performed.

W_{repass}^c can be calculated from the value of i_{pass} deduced from measurement of R_p when the sample is in a steady passive state before applying sliding.

W_{repass}^m can be calculated from Equation 5.6, if W_t, W_{repass}^c and W_{act}^c are known:

- In an intermittent sliding test, EIS measurements cannot be implemented for determining I_{act}, because of the large variations of the current during the stop step. Nevertheless, the value of W_{act}^c for intermittent sliding conditions can be calculated from Equation 5.40 from the value of W_{act}^c determined in the experiment under continuous sliding.
- If a tribocorrosion experiment under continuous sliding at lower rotation rate is carried out instead of an intermittent sliding test, then EIS measurements can be implemented and I_{act} can be deduced from the measured value of R_p. This evaluation is based on the application of the R_p technique to a heterogeneous surface where anodic and cathodic areas are macroscopically distant. The validity of this procedure in a geometry such as that of the ball-on-disc tribometer can be validated by numerical simulation (see Chapter 4).

Ex-situ techniques are also needed in this test for observation of the surface after friction and analysis of its topography:

Many techniques of profilometry or 3D microtopography are available to survey the shape of the wear track and quantify the volume of total wear W_t.

Optical microscopy and scanning electron microscopy observations are also useful to highlight some particular features of the wear process and make sure they are not in contradiction with the assumptions used in the analysis of the test results. For example, it is important to ensure that corrosion does not affect areas outside of the wear track. The observation of the morphology of the wear track is also interesting to qualitatively assess the importance of mechanical wear and check that the tribological conditions applied are consistent with those required for the test.

5.7   Conclusions

A test aimed at understanding the various contributions of a passivating metal to tribocorrosion has been proposed on the basis of the concept of an active wear track and of a repassivation function. The main conditions to be fulfilled in terms of tribology and electrochemistry have been formulated. The expressions of the current under the tribological regimes involved in the protocol and their relationships to the components of the balance of material loss are derived. The next chapters are essentially devoted to the implementation of the test and to its application to real-life materials.

References

1. D. Déforge, F. Huet, R. P. Nogueira, P. Ponthiaux and F. Wenger: *Corrosion*, 2006, **62**, 514–521.
2. V. Vignal, P. Ponthiaux and F. Wenger: *Wear*, 2006, **261**, 947–953.
3. N. Diomidis, N. Göçkan, P. Ponthiaux, F. Wenger and J.-P. Celis: *Intermetallics*, 2009, **17**, 930–937.
4. V. A. D. Souza and A. Neville: *Wear*, 2005, **259**, 171–180.
5. 'Standard guide for determining synergism between wear and corrosion'. ASTM G119-93; 529–534; 1998, West Conshohocken, PA, ASTM.
6. F. Assi and H. Böhni: *Wear*, 1999, **233–235**, 505–514.
7. R. J. K. Wood: *Wear*, 2006, **261**, 1012–1023.
8. Y. Zheng, Z. Yao, X. Wei and W. Ke: *Wear*, 1995, **186–187**, 555–561.
9. I. Garcia, D. Drees and J. P. Celis: *Wear*, 2001, **249**, 452–460.
10. N. Diomidis, J.-P. Celis, P. Ponthiaux and F. Wenger: *Lubr. Sci.*, 2009, **21**, 53–67.
11. N. Diomidis, J.-P. Celis, P. Ponthiaux and F. Wenger: *Wear*, 2010, **269**, 93–103.
12. D. Landolt, S. Mischler and M. Stemp: *Electrochim. Acta*, 2001, **46**, 3913–3929.
13. J.-P. Celis, P. Ponthiaux and F. Wenger: *Wear*, 2006, **261**, 939–946.
14. L. Benea, P. Ponthiaux, F. Wenger, J. Galland, D. Hertz and J. Y. Malo: *Wear*, 2004, **256**, 948–953.
15. D. D. Macdonald: *J. Electrochem. Soc.*, 1992, **139**, 3434–3449.
16. M.-G. Vergé, C.-O. A. Olsson and D. Landolt: *Corros. Sci.*, 2004, **46**, 2583–2600.
17. D. D. Macdonald: *J. Electrochem. Soc.*, 1992, **139**, 3434–3449.
18. E.-A. Cho, C.-K. Kim, J.-S. Kim and H.-S. Kwon: *Electrochim. Acta*, 2000, **45**, 1933–1942.
19. P. Jemmely, S. Mischler and D. Landolt: *Wear*, 2000, **237**, 63–76.
20. D. Hamm, K. Ogle, C.-O. A. Olsson, S. Weber and D. Landolt: *Corros. Sci.*, 2002, **44**, 1443–1456.
21. P. Schmutz and D. Landolt: *Electrochim. Acta*, 1999, **45**, 899–911.
22. P. L. Andresen and F. P. Ford: *Corros. Sci.*, 1996, **38**, 1011–1016.
23. B. S. Lee, H. S. Chung, K.-T. Kim, F. P. Ford and P. L. Andersen: *Nucl. Eng. Des.*, 1999, **191**, 157–165.
24. E. Lemaire and M. Le Calvar: *Wear*, 2001, **249**, 338–344.

25. F. Wenger, P. Ponthiaux, L. Benea, J. Peybernès and A. Ambard: in 'Electrochemistry in light water reactors', (ed. R. W. Bosch, D. Féron and J.-P. Celis), Vol. 49, 195–211; 2007 European Federation of Corrosion Publications. Cambridge, Woodhead Publishing and Leeds, Maney Publishing. ISSN 1354-5116.

26. P. Ponthiaux, F. Wenger, D. Drees and J. P. Celis: *Wear*, 2004, **256**, 459–468.

27. D. Déforge: 'Etude de l'usure des aciers inoxydables dans l'eau a differentes temperatures et sous pression. Approche Tribocorrosion'. Thesis, 2006, Ecole Centrale Paris, Châtenay-Malabry, France.

6

Towards a standard test for the determination of synergism in tribocorrosion: Design of a protocol for passivating materials

Nikitas Diomidis

Ecole Polytechnique Fédérale de Lausanne, Lausanne, Switzerland

ndiomidis@yahoo.com

Existing approaches to analysing the synergism between corrosion and wear in tribocorrosion are reviewed, and their drawbacks are listed. A novel approach to that synergism for passivating materials is presented in this chapter. The novelty of this approach is the fact that the wear track evolves with time in a cyclic way from a passive to an active surface, and the reverse is also taken into account. The galvanic coupling between different surface states generated on passivating materials in sliding contacts is also addressed. A number of relationships are derived allowing researchers to express the synergism in tribocorrosion.

6.1 Synergy in tribocorrosion: previous approaches

One of the early definitions of the synergistic effect between corrosion and wear was introduced by Watson *et al.* [1]. In that work, the authors expressed the total material loss under tribocorrosion, W_t, as the sum of material loss due to pure wear, W^m, plus the material loss due to pure corrosion, W^c, and the material loss due to the synergy between wear and corrosion, W^s, as expressed in Equation 6.1.

$$W_t = W^m + W^c + W^s \qquad [6.1]$$

The wear–corrosion synergism component was further divided as:

$$W^s = W^{cm} + W^{mc} \qquad [6.2]$$

with W^{cm} is the increase in mechanical wear due to corrosion, and W^{mc} the increase in corrosion due to mechanical wear. This concept was further developed by Stack and co-workers for mapping [2] and modelling [3] of the tribocorrosion process. This became a quite widely adopted formalism in the description of the synergistic effect between corrosion and wear, and is most commonly expressed by Equation 6.3

$$W_t = W^{co} + W^{mo} + W^c + W^m \qquad [6.3]$$

with W^{co} the material loss due to corrosion in the absence of wear; W^{mo} the wear in the absence of corrosion; W^c the wear-induced corrosion, and W^m the corrosion-induced wear. Such an approach is used in the current ASTM G119-09 'Standard guide for determining synergism between wear and corrosion' [4].

Despite the fact that this approach emphasises the interdependence between wear and corrosion, it has received criticism in the scientific literature [5, 6]. The main shortcoming resides in the fact that until now there are no experimental methodologies that allow the separate measurement of the different contributions to the total wear loss under tribocorrosion. To overcome this shortcoming, researchers have used quite arbitrary experimental standards for the definition of each contribution. As an example, in the ASTM standard G119, the material loss due to wear in the absence of corrosion is measured by performing wear tests under cathodic polarisation at –1 V with respect to the open circuit potential. Indeed, under such test conditions, corrosion can be assumed to be negligible, and all of the material loss noted can be ascribed to wear. The problem with cathodic polarisation resides however in the absence of corrosion products whose presence can greatly affect the mechanical response of tested surfaces. As a consequence, this measured material loss due solely to mechanical interactions does not necessarily represent correctly the real-life wear behaviour of the tested materials under open circuit potential conditions. Indeed, it is known that the surface state of materials affects tribological properties such as friction, wear, and lubrication. Surface films such as oxides, hydroxides or even contaminants adsorbed from the environment, act on the chemical component of friction, for example, by modification of the surface energy or changes in short distance attraction/repulsion van der Waals forces.

The importance of the presence of corrosion products on the mechanical response of sliding surfaces is seen when performing unidirectional sliding tests at different latency times. Latency time, t_{lat}, is defined as the time between two successive contacts at a given point on the sliding track. This effect is generally not scored when the sliding velocity is large, namely when the contact frequency is high compared to the electrochemical process taking place at the sliding surfaces. On the other hand, in sliding tests performed at a low contact frequency, the importance of the presence of corrosion products on the mechanical response of sliding surfaces is reflected in the dependence of the coefficient of friction in unidirectional sliding tests on the latency time. Since the growth of a passive surface film takes place during this latency time, it can be stated that each level of passive film restoration can be associated with a specific value of the coefficient of friction. This is illustrated in Figure 6.1 for the case of an aluminium-based alloy immersed in a phosphate buffer solution.

Additionally, under cathodic polarisation, a hydrogen embrittlement is a realistic risk, particularly for passive metals [8]. The amount of hydrogen embrittlement usually depends on the applied cathodic potential, and affects the wear volume. As a result, the material loss under cathodic protection cannot be considered as an absolute reference for 'wear in the absence of corrosion'. Furthermore, such an approach does not take into account the galvanic coupling between the worn (depassivated) and unworn (passive) areas on the test sample, and which affects the corrosion rate. A further shortcoming of this approach is that it considers the wear track as a whole, and thus does not provide any local insight into the processes taking place in different parts of the wear track [9].

The above-mentioned shortcomings of the previous approaches to synergy in tribocorrosion support the need for a new approach that provides a deeper scientific insight into the physical phenomena taking place at a tribological contact operating in a corrosive environment. This new approach should also take into account the latest scientific advances in tribology and electrochemistry. Furthermore, such a new approach should provide the practicing engineer with a useful measurement methodology that allows a quantitative estimation of the material loss, a tool for the characterisation and screening of candidate materials for specific applications, and also a way to identify the source of problems arising from a material loss, and thus hinting at possible solutions.

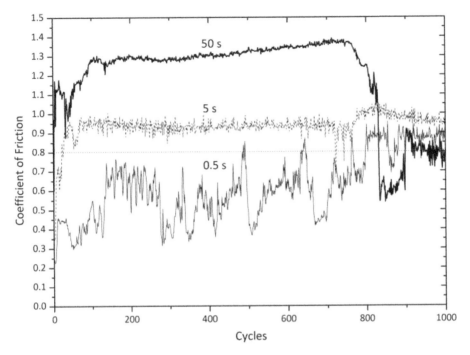

6.1 Evolution of the coefficient of friction of $Al_{71}Cu_{10}Fe_9Cr_{10}$ immersed in a phosphate buffer solution, pH 7 with the number of sliding cycles for three latency times: t_{lat} = 0.5 s (corresponding to unidirectional sliding at t_{rot}= 0.5 s), t_{lat} = 5 s and t_{lat} = 50 s (both corresponding to intermittent unidirectional sliding). The horizontal dotted line corresponds to the coefficient of friction recorded under similar but dry unidirectional sliding at t_{rot}= 0.5 s [7]

6.2   Synergy in tribocorrosion: design of a new test protocol

This new approach to synergy in tribocorrosion is based on the fact that the surface state of a wear track evolves with time in a cyclic manner. Apart from the general wear to the surface, which is common for all tribological conditions, in the specific case of tribocorrosion of passivating materials, another kind of evolution also takes place. This evolution is due to the repeated removal and subsequent re-growth of a passive surface film when a mechanical loading is applied. By controlling the frequency of such depassivation–repassivation events with respect to the time necessary for film growth, it is possible to measure the properties of the surface at different stages of activity and repassivation. To perform tests with different values of latency time, t_{lat}, two approaches are available, but each approach has advantages and limitations as detailed in Chapter 5:

–   *first approach*: Under *continuous sliding* tests, the latency time t_{lat} can be modified by changing the rotation period t_{rot}. Another way to obtain the same t_{lat} is to keep the sliding speed constant and increase the radius of the track. However, for practical reasons, it is not realistic to consider increasing the radius of the track by a large factor.

–   **second approach**: The latency time is changed by performing *intermittent sliding tests*. During such an intermittent sliding test, the counter body slides for one cycle, and then stays immobile for a certain period of time to allow a part of the passive film to re-grow. Thus, an off-time, t_{off}, is imposed at the end of each cycle. As a result, the latency time between two subsequent contact events, t_{lat}, differs from the rotation period, t_{rot}:

$$t_{lat} = t_{rot} + t_{off}$$ [6.4]

Clearly, in continuous sliding tests, Equation 6.4 is still valid but t_{off} is zero.

Thus, that approach should result in a protocol that provides information on the evolution of the surface with testing time, and the identification of the resulting mechanisms of material loss and surface degradation. The *synergistic effect between friction and corrosion* or vice versa, is schematically represented in Figure 6.2.

Therefore, this protocol is elaborated as a series of successive steps during which essential data on the tribocorrosion behaviour of passivating materials are acquired. Depending on the outcome of each step, a decision has to be taken on the test conditions in the following steps or the requirements for the understanding of the underlying mechanisms of material degradation. This protocol has the significant benefit that it enables the practicing scientist or engineer to identify the different mechanisms leading to material loss, namely electrochemical and mechanical mechanisms, to quantify their contribution, but also to identify their origin from different parts of the wear track, namely, active areas and fully or partly repassivated ones. Such an approach can then provide a fundamental insight into the phenomena taking place at a sliding contact in the presence of a corrosive environment. Additionally, it constitutes a useful tool for:

–   the analysis of the local conditions of a real-life application
–   the assessment of the dangers of material loss
–   the identification of the source of possible problems, and
–   the screening of candidate materials for a specific purpose.

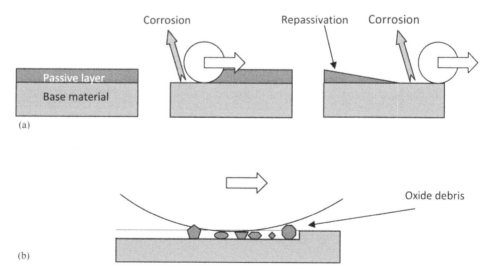

6.2 Schematic representation of (a) corrosion accelerated by friction, and (b) abrasion accelerated by corrosion products [6]

6.2.1 Selection of test conditions

In order to characterise the sensitivity of the tested material to tribocorrosion or to compare the sensitivity of different materials, a careful selection of the test conditions should be carried out before any testing, so as to obtain useful results.

Selection of environmental conditions:

- The electrolyte should be selected based on its known oxidative or reducing power. In this case, the test temperature can be ambient temperature (20–25°C). The test electrolyte should be used at a constant temperature because temperature affects the reaction kinetics at material surfaces (chemical, physical and electrochemical).
- A decisive characteristic of the test electrolyte is its pH. The selection of the pH of the test electrolyte can be based on Pourbaix diagrams [10] (for example, see Figure 6.3 for iron and chromium). Such diagrams are calculated from thermodynamic data and give potential–pH domains where a metal is passive, corrodes or remains immune. In the case of a metallic alloy consisting of a number of elements, the pH range, where at least one of the constituents passivates, should be selected by preference as a test condition. An electrolyte composition that may cause localised corrosion (pitting corrosion in particular) should be avoided.
- A significant change in the composition or pH of the test solution during the test would result 'de facto' in a modification of the electrochemical conditions at the interface between metal and solution. As a result, it is reasonable to provide an amount of solution of at least 50 cm^3 per square centimetre of immersed area to limit the pollution of the solution by the products of corrosion or wear during the test.
- To carry out electrochemical measurements under suitable conditions, avoiding thin layer effects, an electrolyte height of about 1 cm above the immersed surface should be provided.
- In tests at different sliding velocities, a flow or stirring of the electrolyte could be implemented to limit the effects of hydrodynamic conditions generated by sliding.

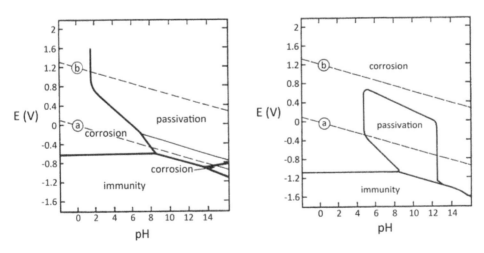

6.3 Schematic E–pH Pourbaix diagrams for iron (left) and chromium (right) [10]. The lines (a) and (b) delineate the E–pH stability range of water

Furthermore, parameters linked to the sliding tests need to be selected by following the recommendations hereafter:

- Normal force, F_n: The value of the normal force should be selected so as to avoid plastic deformation of the bulk material. The normal force should be selected so that the maximum Hertzian contact pressure on the test material before starting sliding is smaller than the yield strength. When sliding is activated on non-conformal contacting surfaces, wear occurs and the contact area between the surfaces increases, leading to a gradual decrease in the contact pressure with test duration.
- Track radius, R_{tr}: This should be selected in such a way that edge effects are avoided, for example, in the case of a disc, by preference about half the radius.
- Number of cycles, N: This depends on the type of material to be tested and the test conditions. It should be selected so that the wear volume is large enough to be measured accurately, while avoiding too long test durations for practical reasons. A preliminary sliding test might be necessary to determine N. In the case of coated materials, the maximum wear depth allowable is a fraction of the coating thickness (typically one-tenth as recommended in hardness measurements to avoid effects of the substrate material on the experimental data).

In some particular cases, the selection of test conditions can also be done so as to reflect the behaviour of the test material during a real-life application (see also in Chapter 3, Section 3.4.1 on the TAN approach).

6.2.2   Electrochemical tests on passive material without any sliding

After selecting appropriate test conditions, measurements are performed to collect information on the properties of a material when fully covered by a passive film. This is done by electrochemical tests in the absence of any sliding. The material is immersed in the test electrolyte, and the open circuit potential, E_{oc}, is measured versus a reference electrode. In general, a stable value of E_{oc} is obtained after some interval of immersion. From an electrochemical point of view, a stable value of E_{oc} is obtained when the long-term fluctuations of E_{oc} are below 1 mV min^{-1} for a minimum of 1 h. The time necessary to reach such a stationary open circuit potential in the test electrolyte is an important characteristic of a passivating process, and in this proto-col, is called the *reaction time* characteristic, t_{reac}. A stable E_{oc} indicates that dynamic equilibrium is achieved between the surface of a material and its environment. The evolution of E_{oc} from immersion time onwards provides useful information on the electrochemical reactivity of the tested material in the test electrolyte (see Figure 6.4):

From Figure 6.4, one can derive that when:

- E_{oc} decreases with time, general corrosion may be suspected
- there are short-term potential fluctuations, then there is a risk of localised corrosion such as pitting, or periodic adsorption/desorption processes. Often, pitting occurs after a first step of 'incubation', where the potential increases monotoni-cally
- E_{oc} increases with time, passivation or adsorption is probably taking place.

In this latter case, t_{reac} can be estimated from the evolution of E_{oc} with immersion time by drawing the tangent to the curve at the point where the slope is maximum,

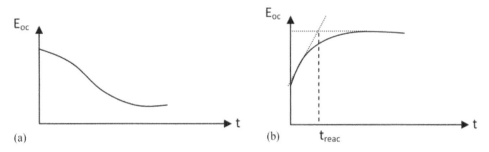

6.4 Schematic representation of the evolution of E_{oc} with immersion time in the case of (a) corrosion, and (b) passivation. The graphical determination of t_{reac} in the case of passivation is shown in (b)

together with a straight line tangential to the data in the part of the curve where E_{oc} is stable.

After achieving a long-term stable open circuit potential indicative of passivation, the polarisation resistance of the passive material, R_p, is measured by electrochemical impedance spectroscopy (EIS). Based on R_p, the specific polarisation resistance of passive material, r_{pass}, can be calculated for a test sample with a surface area, A_o, as:

$$r_{pass} = R_p. A_o \qquad [6.5]$$

Specific polarisation resistance values for metallic materials of 10^3 Ω cm^2 or lower indicate the presence of an active sample surface, while values around 100×10^3 Ω cm^2 or higher indicate a passive sample surface. The corrosion current density of the material covered by a passive surface film, i_{pass}, is then calculated as follows:

$$i_{pass} = \frac{B}{r_{pass}} \qquad [6.6]$$

with B a constant. For metallic materials, B normally varies between 13 and 35 mV, depending on the nature of the material and the environment. In this protocol, a value of 24 mV is assumed. This current i_{pass} is considered to correspond to the dissolution current of the material through the passive film in a stationary state.

6.2.3   Electrochemical tests on fully active sliding track during sliding tests

The next step is the determination of the corrosion rate of the depassivated material. In order to keep a part of the immersed sample surface in a continuous active state, the passive film has to be removed by mechanical contact. It is thus necessary to select a rotation period, t_{rot}, which is small compared to t_{reac} so that the passive film has no time to re-grow in between two successive contact events. Generally, the rotation period t_{rot} is taken as:

$$t_{rot} = \frac{t_{reac}}{10000} \qquad [6.7]$$

Considering now the case in which t_{rot} as expressed in Equation 6.7 is achieved, a first t_{latl} is then taken equal to t_{rot}. The surface state of the disc is then as schematically shown in Figure 6.5 in the case of a ball-on-disc test.

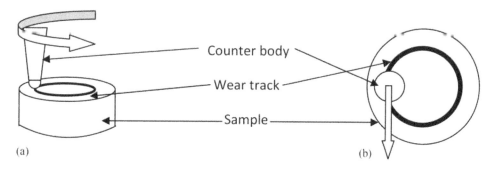

6.5 A schematic representation of the experimental set-up used for unidirectional sliding tests: (a) front view, and (b) top view [6]

It is thus assumed that during such sliding tests, the whole wear track area is in an active state, so that:

$$A_{tr} = A_{act}$$
[6.8]

The different dimensions of ball and wear track on the disc are schematically shown in Figure 6.6.

Despite the fact that the width of the sliding track increases progressively due to wear during sliding tests performed against a curved counter body, a mean sliding track area, A_{tr}, is taken for simplicity, and is defined as:

$$A_{tr} = \frac{1}{2}\left(A_{tr\,max} + A_{tr\,min}\right)$$
[6.9]

with $A_{tr\,max}$ the maximum value measured at the end of the test, and $A_{tr\,min}$ the minimum value at the end of the first cycle. $A_{tr\,min}$ is calculated by multiplying the length of the wear track, L, by the diameter of the Hertzian static contact area, e:

$$e = 2\left(\frac{3\,F_n R}{4\,E}\right)^{1/3}$$
[6.10]

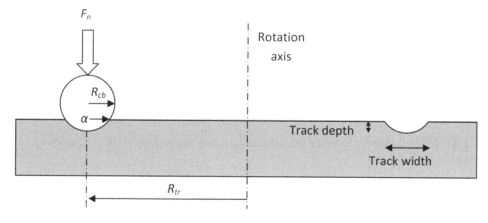

6.6 A magnified representation of the experimental set-up used for unidirectional sliding tests [6]

with F_n the applied normal load, R the radius of the tip of the curved counter body, and E the equivalent elastic modulus given by:

$$\frac{1}{E} = \frac{1-v_1^2}{E_1} + \frac{1-v_2^2}{E_2} \qquad [6.11]$$

with v_1 and v_2 the Poisson's ratios, and E_1 and E_2, the elastic moduli of the test sample and counter body, respectively.

The E_{oc} of the test sample is monitored starting at immersion. The sliding is initiated at the time a stable E_{oc} is achieved. The E_{oc} value recorded during sliding is a mixed potential resulting from the galvanic coupling of two types of material present on the sample surface, namely the material inside (A_{tr}) and outside (A_o–A_{tr}) the sliding track. It is assumed that the kinetics of the redox reactions taking place on each of these areas, do not vary with the real potential of the sliding track. In other words, the ohmic drop effect is considered to be negligible in the galvanic coupling between the sliding track and the surrounding area.

EIS measurements are performed during sliding to measure the polarisation resistance, R_{ps}, of the sample surface. Similarly to E_{oc}, R_{ps} may be considered as the combination of two polarisation resistances, namely R_{act} related to the active area A_{act} which is equal in this case to the wear track, and R_{pass} which corresponds to the surrounding unworn area, (A_o–A_{tr}):

$$\frac{1}{R_{ps}} = \frac{1}{R_{act}} + \frac{1}{R_{pass}} \qquad [6.12]$$

where:

$$R_{act} = \frac{r_{act}}{A_{tr}} \qquad [6.13]$$

and:

$$R_{pass} = \frac{r_{pass}}{A_0 - A_{tr}} \qquad [6.14]$$

Since r_{pass}, is known from Equation 6.5, it is possible to calculate the specific polarisation resistance of the active surface by substituting Equations 6.11 and 6.1 into Equation 6.10:

$$r_{act} = \frac{A_{tr}\, R_{ps}\, r_{pass}}{r_{pass} - R_{ps}\left(A_0 - A_{tr}\right)} \qquad [6.15]$$

It is now possible to calculate the corrosion current density of the active material, i_{act}, by substituting r_{pass} with r_{act} in Equation 6.6:

$$i_{act} = \frac{B}{r_{act}} \qquad [6.16]$$

6.2.4   Electrochemical tests on partially active sliding track during sliding tests

In the preceding steps of the protocol, two extreme cases were characterised, i.e. the passive material and the active one. Under tribocorrosion conditions at high latency times, the surface of the material undergoes sequential events of depassivation and repassivation in-between successive contacts. This means that a part of the surface at any given time repassivates progressively. The latency time is then selected so that the

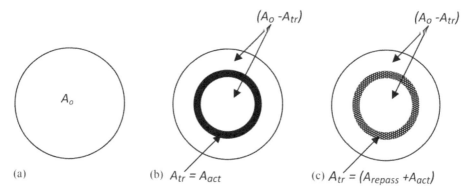

(a)

(b) $A_{tr} = A_{act}$

(c) $A_{tr} = (A_{repass} + A_{act})$

6.7 Schematic top view of a disc shaped test sample: (a) without sliding, (b) during sliding tests at latency times resulting in a fully active sliding track, and (c) during sliding tests at latency times resulting in a partially active sliding track. Active material area is shown in black while passive and repassivated material areas are shown in white [6]

re-growth of a surface film between two successive contact events is not negligible anymore as was the case under sliding at low latency times. The different surface states on the disc are shown in Figure 6.7 for three different sliding test conditions.

To achieve partially active sliding tracks, it may be recommended to select $t_{lat2} = t_{reac}/1000$ and $t_{lat3} = t_{reac}/100$.

As a result of the increase in the latency time in this step, the wear track can be assumed to consist of two distinct zones [9, 11, 12], namely:

- a fraction of the sliding track from which the initial passive film has been removed during sliding. In this area, the test material may be either bare, or covered by a reaction layer different from the initial passive film. This area is named the active area, A_{act}, and
- the remaining sliding track area covered by a surface film that is in the same state as the surface before sliding. This film is either not removed by the counter body during sliding, or it had time to restore its initial state. This area is referred to as the repassivated area, A_{repass}, with:

$$A_{tr} = A_{act} + A_{repass} \qquad [6.17]$$

It must be noted that under continuous sliding, these active and repassivated areas remain constant because of stationary electrochemical state conditions. Under intermittent sliding, these active and repassivated areas on the sliding track evolve with time between two successive contact events since a gradual increase in the coverage of the repassivated area takes place within the *off* period. By hypothesis, in both cases, the fraction of the sliding track surface covered by the passive film, A_{repass}/A_{tr}, is assumed to be constant and given by the ratio t_{lat}/t_{reac}:

$$\frac{A_{repass}}{A_{tr}} = \frac{t_{lat}}{t_{reac}} \qquad [6.18]$$

and:

$$\frac{A_{act}}{A_{tr}} = 1 - \frac{t_{lat}}{t_{reac}} \qquad [6.19]$$

As a result, for the proposed values of the latency time $t_{lat2} = 0.001\ t_{reac}$ and $t_{lat3} = 0.01\ t_{reac}$, the relationships between the repassivated area and the total wear track area are, respectively, $A_{repass\ 2} = 0.001\ A_{tr}$ and $A_{repass\ 3} = 0.01\ A_{tr}$, and thus $A_{act\ 2} = 0.999\ A_{tr}$ and $A_{act\ 3} = 0.99\ A_{tr}$.

6.3   Analysis and interpretation of sliding test results

The procedure for analysing the results developed below is only applicable to tests performed with a constant track radius. For tests with a varying track radius, the procedure for analysing the results should take account of the effect on the wear resulting from the increase in the wear track area (A_{tr}).

As proposed in the previous chapter, we arbitrarily chose to express the total wear W_{tr} (measured in the trace of friction of area A_{tr}) as a sum of components related to both types of area present on the track, the area A_{act} in the active state, and the area A_{repass} in the repassivated state:

$$W_{tr} = W^c_{act} + W^m_{act} + W^c_{repass} + W^m_{repass} \qquad [6.20]$$

with:

- W_{tr} the material loss in wear track
- W^c_{act} the material loss due to corrosion of active material in wear track
- W^m_{act} the material loss due to mechanical wear of active material in wear track
- W^c_{repass} the material loss by corrosion of repassivated material in wear track
- W^m_{repass} the material loss due to mechanical wear of repassivated material in the wear track.

The analysis of the test results will be carried out to assess the values of these different components and to compare them to determine the characteristics of the wear mechanism.

6.3.1   Corrosion and mechanical wear of active material at low latency times

Under continuous sliding at low latency times $t_{lat\ 1}$, the calculation of the corrosion current density $i_{act\ 1}$ of the active material was explained in Section 6.2.3 and deduced from polarisation measurement by EIS. The volumetric material loss due to corrosion of the active surface, $W^c_{act\ 1}$, can be calculated for sliding tests at low latency times using the appropriate A_{act} and t_{lat} values ($A_{act} = A_{act\ 1}$ and $t_{lat} = t_{lat\ 1}$) in Equation 6.21:

$$W^c_{act\ j} = \left(\frac{C_F}{Fd}\right) i_{act\ j}\ A_{act\ j}\ N\ t_{lat\ j} \qquad [6.21]$$

with $j = 1$, d the density of the test material, and F the Faraday constant (96 485 C mol^{-1}). In the case of a pure metal, C_F is given by:

$$C_F = \frac{M}{n} \qquad [6.22]$$

with M the atomic weight, and n the number of electrons involved in the oxidation reaction. In the case of alloys containing z alloying elements, and assuming that the different elements i oxidise in proportion to their weight content x_i in the alloy, C_F can be calculated as:

$$C_{i}^{-1} = \sum_{i=1}^{i=z} \frac{x_i \, n_i}{M_i} \qquad [6.23]$$

with M_i the atomic weight of element i_i and n_i the number of electrons involved in the oxidation reaction of element i.

The volumetric material loss due to mechanical wear of the active material, W^m_{act}, during a sliding test at low latency time can be calculated by subtracting the material loss due to corrosion, W^c_{act}, from the total material loss in the wear track, W_{tr}, measured after the end of the sliding test:

$$W^m_{act\,1} = W_{tr} - W^c_{act\,1} \qquad [6.24]$$

6.3.2 Corrosion and mechanical wear of active material at high latency times

For tests performed at high latency times (t_{lat2} and t_{lat3}), the processing of the results for calculating W^c_{act} depends on the chosen approach (Section 6.2.4):

- **First approach:** Under continuous sliding tests, the calculation of the corrosion current density i_{act} of the active material can be performed as explained in Section 6.2.3 and deduced from polarisation resistance measurement by EIS. Then, the volumetric material loss due to corrosion of the active surface, W^c_{act}, can be calculated for sliding tests at high latency times using the appropriate A_{act} and t_{lat} values in Equation 6.21.
- **Second approach:** Under intermittent sliding tests, EIS measurements cannot be performed as was explained in Section 6.2.3 due to the instability of E_{oc}. The value of i_{act} cannot be determined. An alternative solution is to use Equation 5.39 defined in Chapter 5 using the appropriate A_{act} and t_{lat} values in Equation 6.25:

$$W^c_{act\,j} = W^c_{act1} \left(\frac{t_{lat\,j}}{t_{lat1}} \right)^{(1-p)} \frac{A_{act\,j}}{A_{act\,1}} \qquad [6.25]$$

with $j = 2$ or 3, and W^c_{act1} and t_{lat1} the corrosive wear component and the latency time, respectively, related to the continuous sliding test at low latency times (Section 6.2.3).

Regardless of the approach implemented, to calculate W^m_{act} for the case of sliding tests at high latency times, it is assumed that the duration of the latency time does not affect the mechanical wear resistance of the active material. Consequently, $W^m_{act\,j}$ for a latency time $t_{lat\,j}$ is calculated from the value $W^m_{act\,1}$ obtained for a latency time $t_{lat\,1}$ by the following equation:

$$W^m_{act\,j} = W^m_{act\,1} \left(\frac{A_{act\,j}}{A_{act\,1}} \right) \qquad [6.26]$$

with $j = 2$ or 3.

6.3.3 Corrosion and mechanical wear of the repassivated material in the wear track

The corrosion and mechanical wear on the repassivated area A_{repass} is derived solely from sliding tests at high latency times since they allow the study of the periodic removal and re-growth of a passive surface film. The material loss due to corrosion of the repassivated area of the wear track, W^c_{repass}, is calculated by analogy to Equation 6.17 but for a repassivated material as:

$$W^c_{repass\ j} = \left(\frac{C}{Fd}\right) i_{pass}\ A_{repass\ j}\ N\ t_{lat\ j} \qquad [6.27]$$

with $j = 2$ or 3, since it is assumed that the repassivated material in A_{repass} has the same corrosion behaviour as the starting passive material in A_o. The material loss in the sliding track due to the mechanical wear on the passive area, W^m_{repass}, is then obtained from:

$$W^m_{repass} = W_{tr} - \left(W^c_{repass} + W^c_{act} + W^m_{act}\right) \qquad [6.28]$$

It must be noted that, depending on the wear mechanism involved, the term $W^m_{repass\ j}$ corresponds either to the wear of the passive film alone, or to the wear of the film and of a part of the underlying material.

6.3.4   Interpretation of the sliding test results

A more detailed assessment of the resistance of a material to tribocorrosion is possible based on the parameter K_c being the ratio between corrosive and mechanical material losses, namely

$$K_c = \frac{W^c_{act} + W^c_{repass}}{W^m_{act} + W^m_{repass}} \qquad [6.29]$$

The following three cases can be distinguished based on the value of K_c:

- when $K_c > 1$, corrosion is the predominant contribution to material loss. The total wear will be mainly controlled by the reactivity of the substrate in the test environment
- when $K_c < 1$, the material loss due to mechanical removal predominates, and
- when $K_c \ll 1$, the contribution due to the acceleration of corrosion induced by the destruction of the passive film, even though it is large ($W^c_{act} \gg W^c_{repass}$), will be negligible compared to the total wear.

The effect of the formation of the passive film on the mechanical wear can be evaluated based on K_m, the ratio between the specific mechanical wear on the active and on the repassivated parts of the sliding track, W^m_{act}/A_{act} and W^m_{repass}/A_{repass}, respectively:

$$K_m = \frac{W^m_{act\ j}}{W^m_{repass\ j}} \frac{A_{repass\ j}}{A_{act\ j}} \qquad [6.30]$$

with $j = 2$ or 3.

The following two cases can be distinguished based on the value of K_m:

- when $K_m > 1$, the passive film provides a protection to the material against mechanical removal. The material will be more sensitive to tribocorrosion at short latency times, and
- when $K_m < 1$, the formation of the passive film accelerates the mechanical removal of the material. The sensitivity to tribocorrosion will increase at increasing latency times. If a passive film has the time to re-grow, the mechanical sliding will remove a larger amount of material than when the passive film was not present.

These additional parameters, K_c and K_m, allow a new description of the tribocorrosion regime to be developed compared to previous modelling attempts [13 15]. With these parameters, the synergy between electrochemical and mechanical effects is de-convoluted for the first time into specific components originating from the different zones in the wear track. Additionally, the importance of the different mechanisms leading to material degradation (mechanical or electrochemical ones) can be assessed. This new approach to tribocorrosion provides a detailed insight into the phenomena taking place at a sliding contact immersed in a corrosive environment. Additionally, it may be used as a tool by engineers to analyse the local conditions of a real life application, to assess the dangers of material loss, to identify the sources of possible problems, and thus to screen for the most suited candidate materials.

6.4   Comparison between previous and this new approach of synergism in tribocorrosion

In the new approach to synergy in tribocorrosion presented earlier, the total wear loss is decomposed as expressed by Equation 6.20 into components directly related to the two kinds of areas in the wear track, namely the area A_{act} where the material is in the active state, and the area A_{repass} where the material is covered with a passive film.

In the literature, a different kind of decomposition of the total wear loss under tribocorrosion was presented as expressed by Equation 6.3. A correlation between these two approaches could result in a link between the active wear track concept of the new approach and the synergy concept of the old approach. This can also provide a new interpretation of the synergy between wear and corrosion.

Let us define the specific material loss components in the sliding track, w^j_n, per unit area, based on the material loss and the corresponding surface area:

$$w^j_n = \frac{W^j_n}{A_n} \qquad [6.31]$$

with $n =$ act or repass and $j =$ c or m.

The value w^j_n corresponds to the thickness of the material removed by either corrosive or mechanical action from active or repassivated parts of the sliding track.

By combining Equations 6.17, 6.31 and 6.20, the expression of W_{tr} becomes:

$$W_{tr} = w^c_{act} A_{act} + w^m_{act} \left(A_{tr} - A_{repass} \right) + w^c_{repass} \left(A_{tr} - A_{act} \right) + w^m_{repass} A_{repass} \qquad [6.32]$$

or:

$$W_{tr} = w^c_{repass} A_{tr} + w^m_{act} A_{tr} + \left(w^c_{act} - w^c_{repass} \right) A_{act} + \left(w^m_{repass} - w^m_{act} \right) A_{repass} \qquad [6.33]$$

The first term, $w^c_{repass} A_{tr}$, in Equation 6.33 represents the wear due to corrosion of an area A_{tr} corresponding to the surface of the wear track covered by a passive film. This is thus the wear track in contact with the environment in the absence of sliding as represented by the term W^{co} in Equation 6.3:

$$W^{co} = w^c_{repass} A_{tr} \qquad [6.34]$$

In the same manner, the term $w^m_{act} A_{tr}$ represents the mechanical wear in the wear track not covered by a passive film. This is thus the mechanical wear in the absence of the effect of the environment as represented by the term W^{mo} in Equation 6.3:

$$W^{mo} = w^m_{act} A_{tr} \qquad [6.35]$$

The third term of Equation 6.33, namely $(w^c_{act}-w^c_{repass})A_{act}$, corresponds to the increase (or decrease) of the corrosive component of material loss resulting from the presence of an area depassivated by sliding, namely A_{act}. This corresponds thus to the increase in the specific corrosive wear, $(w^c_{act}-w^c_{repass})$, resulting from the transformation of a repassivated surface into a depassivated one. This term is represented in Equation 6.3 by W^c, and expresses the evolution of the corrosive component of the material loss due to the mechanical perturbation of sliding (effect of sliding on corrosion). It is thus:

$$W^c = \left(w^c_{act} - w^c_{repass}\right)A_{act} \qquad [6.36]$$

The final term $(w^m_{repass}-w^m_{act})A_{repass}$ corresponds to the increase or decrease in the mechanical component of material loss due to the presence of the passive film in the area A_{repass}. This corresponds to the increase in the specific mechanical wear provoked by the restoration of the passive film. This term corresponds in Equation 6.3 to W^m, and expresses the evolution of the mechanical component of the material loss due to the presence of the environment (effect of corrosion on the mechanical wear). It is thus:

$$W^m = \left(w^m_{repass} - w^m_{act}\right)A_{repass} \qquad [6.37]$$

The relationships established above between the different expressions of the material loss due to tribocorrosion, namely Equations 6.3 and 6.33, allow the interpretation of the results obtained by the new protocol in terms of synergy linking these results to the classical approach. It is thus possible to compare the results obtained by the protocol with those existing in the literature obtained by different methodologies aimed to evaluate the synergy terms. They also allow a new interpretation of the terms of synergy based on the relationship between active and passive areas, and on the variation of the specific wear components.

This new interpretation can not only provide further insight on tribocorrosion under sliding but can also be of interest in other domains, such as erosion–corrosion, where the relationship between active and passive areas is not yet understood.

Additional electrochemical techniques such as polarisation measurements and potential jumps can be implemented to investigate the characteristics of the repassivation reactions taking place in tribocorrosion.

6.5   Galvanic coupling: effect of contact area on open circuit potential

The implementation of this protocol requires a special attention to the effect of the ratio of the worn and unworn areas. This ratio determines the magnitude of the galvanic coupling, and is illustrated through the following case.

In order to reveal any effect of the contact area on the open circuit potential value during sliding, the counter body radius and the normal load were varied simultaneously so that, under two sets of test conditions, an identical maximum contact pressure of 281 MPa was achieved but for two different Hertzian contact areas (namely 0.21 mm^2 and 0.0005 mm^2). The difference in the size of the sliding track area appeared clearly from SEM micrographs of the sliding tracks taken after the sliding tests (see Figure 6.8).

The width of the sliding track obtained in the test performed with an initial Hertzian contact area of 0.21 mm^2 is 0.63 mm. This is four times the width of the sliding track (0.15 mm) obtained with an initial Hertzian contact area of 0.0005 mm^2. The corresponding values for volume loss of each sliding track area at the end of the test are namely and 9×10^{-2} mm^3 and 3×10^{-3} mm^3.

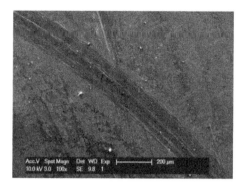

6.8 Scanning electron micrographs of stainless steel after continuous unidirectional sliding in 0.5 M sulphuric acid at E_{oc} performed for 2700 cycles under an initial Hertzian contact area of 0.0005 mm^2 (left), and 0.21 mm^2 (right) [11]. Counter body: corundum balls

The influence of the area of the wear track on the open circuit potential value under sliding is shown in Figure 6.9:

The fact that an identical contact pressure was achieved in both tests ensures that the difference in open circuit potential noted between the two tests shown in Figure 6.9, is only caused by a varying ratio of active-to-passive area, and not by any mechanical effect such as, for example, an elasto-plastic deformation or a load-induced phase transformation. The higher the ratio, the more cathodic the open circuit potential value. At the end of sliding, an increase in the open circuit potential value is observed in both cases, revealing the repassivation of the sliding track. A faster repassivation rate is obtained for the sample at open circuit potential with the smallest active-to-passive area ratio. The different active-to-passive area ratios achieved in this way, clearly affect the open circuit potential and its variation during intermittent sliding tests.

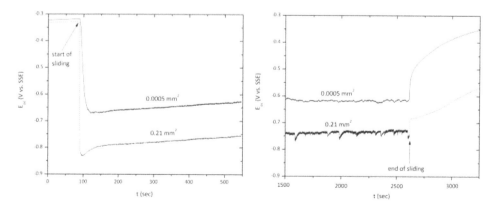

6.9 Evolution of the open circuit potential of AISI 316 stainless steel immersed in 0.5 M sulphuric acid at 25°C, during continuous unidirectional sliding performed under the same Hertzian pressure of 281 MPa but with contact areas of either 0.0005 or 0.21 mm^2. Data shown are at the beginning of the sliding test (left), and at the end of the sliding test (right) [11]. Counter body: corundum balls

As a result, different tests carried out using the protocol can only be compared for ratios of worn area/unworn area of the same order of magnitude, all other test conditions being the same.

6.6   Conclusions

This new protocol seems to be an efficient tool to characterise the relative contributions of mechanical and corrosive components acting on active, and fully or partially repassivated areas in the sliding tracks. This quantitative approach is based on current and potential distributions across different areas of the sliding track and on the kinetics of repassivation. In contrast to the previous approaches proposed for expressing the contribution of the synergy effects to the total wear, the new protocol allows a decomposition of the total material loss in the wear track leading to an estimation of the relative importance of corrosion and mechanical wear of the bare material and the passive film in the wear track. This leads to an identification of the phenomena underlying the degradation of a material, and an assessment of the susceptibility of a passivating material to tribocorrosion. Furthermore, the effect of the environment on the coefficient of friction can be analysed. The application of this protocol allows us to unravel the tribocorrosion mechanisms in a novel way.

The presently proposed approach is a first step. It will be further refined by investigating the effect of galvanic coupling between worn and unworn areas on the kinetics of corrosive wear and on the evolution of the active and repassivated areas with latency time. In fact, concerning the evolution of these parameters, some major and arbitrary assumptions had to be made in the present approach to interpret the experimental results. These assumptions need to be refined or modified as necessary to improve the quantitative validation of the new protocol and to make the test more efficient both in the analysis of the results and in the implementation of the test itself. The improvement in the experimental procedure proposed will surely be of benefit to end-users of this new protocol, and will promote its application in academic and industrial research centres.

References

1. S. W. Watson, F. J. Friedersdorf, B. W. Madsen and S. D. Cramer: *Wear*, 1995, **181–183**, (2), 476–484.
2. M. M. Stack and K. Ch: *Wear*, 2003, **255**, (1–6), 456–465.
3. J. Jiang, M. M. Stack and A. Neville: *Tribol. Int.*, 2002, **35**, (10), 669–679.
4. 'Standard guide for determining synergism between wear and corrosion'. ASTM G119-93, 529–534; 1998, West Conshohocken, PA, ASTM.
5. S. Mischler: *Tribol. Int.*, 2008, **41**, (7), 573–583.
6. N. Diomidis, J.-P. Celis, P. Ponthiaux and F. Wenger: *Lubr. Sci.*, 2009, **21**, (2), 53–67.
7. N. Diomidis, N. Göçkan, P. Ponthiaux, F. Wenger and J.-P. Celis: *Intermetallics*, 2009, **17**, (11), 930–937.
8. S. Akonko, D. Y. Li and M. Ziomek-Moroz: *Tribol. Lett.*, 2005, **18**, (3), 405–410.
9. I. García, D. Drees and J.-P. Celis: *Wear*, 2001, **249**, (5–6), 452–460.
10. M. Pourbaix: 'Atlas of electrochemical equilibria in aqueous solutions'; 1974, Houston, TX, National Association of Corrosion Engineers.
11. N. Diomidis, J.-P. Celis, P. Ponthiaux and F. Wenger: *Wear*, 2010, **269**, (1–2), 93–103.
12. L. Benea *et al.*: *Wear*, 2004, **256**, (9–10), 948–953.
13. J. Jiang and M. M. Stack: *Wear*, 2006, **261**, (9), 954–965.
14. D. Landolt, S. Mischler and M. Stemp: *Electrochim. Acta*, 2001, **46**, (24–25), 3913–3929.
15. D. Landolt, S. Mischler, M. Stemp and S. Barril: *Wear*, 2004, **256**, (5), 517–524.

Towards a standard test for the determination of synergism in tribocorrosion: Detailed testing procedure for passivating materials

Raquel Bayon

TEKNIKER Otaola 20. P.K. 44, SP-20600 EIBAR Gipuzko, Spain

rbayon@tekniker.es

In this chapter, a detailed practical description is given of the different steps foreseen in the protocol on conducting tribocorrosion tests under sliding conditions. It also includes information on sample preparation (surface finish, cleanliness, etc.), testing devices required (potentiostat, reference and counter electrodes), tribological test conditions (applied load, frequency, rotational/oscillatory speed, track size), and electrochemical techniques (open circuit potential measurements, electrochemical impedance spectroscopy conditions). As an illustrative case study, the tribocorrosion components appearing in the protocol are quantitatively calculated for stainless steel coupons immersed in a dilute sulphuric acid solution. The management of the experimental data obtained is discussed.

7.1 Selection of tribocorrosion testing parameters

Tribocorrosion sliding tests are useful to determine the material loss due to the different processes involved. From such tests, the contribution to the material loss from electrochemical effects, mechanical effects, and/or combined electrochemical–mechanical effects, becomes accessible. The most suitable test conditions should be selected specifically for the passive material under study.

The test samples should preferably be polished to a final roughness of around $0.05\,\mu m$ (mirror finish), cleaned with ethanol, air dried, and stored in a desiccator for some time before the start of the tests to allow the formation of a natural surface film. In this way, possible contamination with foreign matter is also prevented.

The test set-up consists of the wear test equipment (tribometer) allowing either rotating ball-on-disc sliding tests or reciprocating ones. The test samples are discs and are fully immersed in the test solution (electrolyte). The counter bodies are loaded on top of the discs and are generally balls or pins with a spherical end. The ball material should be non-conductive to prevent any galvanic coupling with the disc material. A potentiostat is required to perform the electrochemical measurements. A three-electrode set-up has to be installed consisting of a working electrode composed of the polished and cleaned test sample, a counter electrode such as a Pt wire, a Pt plate or a Pt grid, and a reference electrode such as a saturated calomel or silver/silver chloride electrode. The position of the reference electrode is preferably between the wear track and the counter electrode. The risk of distortion of some electrochemical measurements due to a non-uniform distribution of the current lines has to be limited. Therefore, the counter electrode should preferably be in rotating ball-on-disc sliding testers, as a circular electrode positioned with its axis corresponding to the axis of the disc acting as the working electrode. In the case of reciprocating sliding testers, the counter electrode should be flat with a length comparable to the length of the wear track, and positioned parallel to the ball displacement direction.

Tribological parameters have to be chosen as a function of the sample geometry, the mechanical and chemical properties of the tested material, and the limitations of the laboratory devices available. This selection can be made based on the following recommendations:

- *type of contact*: a ball loaded on top of a polished and cleaned flat test sample is recommended
- *counter body*: ceramic balls are used because they are chemically inert. The ball diameter depends on the tribometer capabilities and the desired contact pressure
- *normal load*: this depends on the mechanical properties of the test material, and on the desired average contact pressure. It should also take into account the geometry, dimensions, and mechanical properties of the counter body (see also Chapter 6 Section 6.2.1)
- *contact periodicity*: in the case of rotating ball-on-disc sliding tests, one has to select the rotation speed (t_{rot}) and to decide whether an off time (t_{off}) is used or not. In the case of reciprocating sliding tests, one has to select the sliding frequency and to decide whether an off time (t_{off}) is used. This test parameter is selected by considering the time characteristics of the passivating process, t_{reac}
- *sliding track*: in the case of rotating ball-on-disc sliding tests, one selects the radius of the wear track (R) (see Chapter 6, Section 6.2.1), or the displacement amplitude in the case of reciprocating sliding tests
- *number of contact events*: together with the latency time, this determines the test duration, and
- *test solution* (electrolyte) (see Chapter 6, Section 6.2.1).

7.2   Electrochemical response of the tested material in the absence and presence of sliding resulting in an active sliding track

The testing is divided into four experiments that are performed on the same test sample which is polished and cleaned (see Section 7.1) before initiating the first experiment. In order to obtain information on the repeatability of the experiments, it is strongly recommended to perform at least three repeat tests of each experiment.

7.2.1   Electrochemical impedance spectroscopy at open circuit potential (E_{oc}) in the electrolyte without any sliding (Experiment A)

Objective: To gather information on the surface properties of the passive base material in the absence of any sliding.

Tests to be performed: The cleaned test sample is assembled in the test cell as soon as possible after removal from the desiccator. Avoid any contact of the polished surface with fingers or foreign material. The test cell is subsequently filled with fresh test electrolyte stabilised at room temperature. The following steps have to be performed:

- *Step A.1*: Immerse the sample in the electrolyte and immediately start measuring the open circuit potential of the test sample (E_{oc}) until a stable E_{oc} value is obtained. A stable E_{oc} value is obtained when the long-term fluctuations of E_{oc} are below 1 mV min⁻¹ for a minimum of 1 h. Raise the stable E_{oc} value and determine the value of t_{reac}.

- *Step A.2:* Perform electrochemical impedance spectroscopic measurements at an ac amplitude of 10 mV superimposed on E_{oc}. The starting frequency is 10^5 Hz. Proceed then with measurements performed at frequencies down to at least 10^{-2} Hz. Record data at 10 frequencies per decade.

Report the stable E_{oc} measured, the value of t_{reac}, and the impedance response at the different frequencies.

7.2.2 Evolution of open circuit potential in the electrolyte just before, during, and after sliding generating an active material surface (Experiment B)

Objective: To obtain information on the passive–active surface conditions induced by sliding performed under conditions of short latency time.

Tests to be performed: Rotating or reciprocating sliding tests are initiated immediately after the end of Experiment A.2 without removing the test sample from the electrolyte. The following steps have to be performed:

- *Step B.1*: Proceed with measuring E_{oc} of the test sample for another 2 min.
- *Step B.2:* Load the counter body on the test sample while recording E_{oc} further.
- *Step B.3*: Start the first sliding test under sliding test conditions generating an active material surface in the sliding track. Before this, the latency time and t_{rot} must be selected so that $t_{lat\,1} = t_{rot} = t_{reac}/10\,000$. In the case of reciprocating sliding tests, one has to select the sliding frequency so that the highest exposure time is at maximum equal to $t_{reac}/10\,000$. The other sliding test parameters are selected as under Section 7.1. During sliding, the coefficient of friction, μ, can be recorded too. Record on-line the open circuit potential under sliding, E^s_{oc}, which is a mixed potential between the passive material outside the sliding track and active material inside the sliding track.
- *Step B.4*: Stop the sliding test, but do not remove the test sample from the electrolyte.
- *Step B.5*: Record the open circuit potential, E_{oc} during a 10 min period after the end of sliding. At the end of this 10 min period, do not remove the test sample from the electrolyte but proceed quickly to Experiment C.

After Experiment B.5, report the E_{oc} value recorded just before starting the sliding, the mean value of E^s_{oc} recorded during the sliding period, and the E_{oc} value recorded 10 min after stopping sliding.

7.2.3 Electrochemical impedance spectroscopy under sliding at a fixed potential corresponding to E^s_{oc} (Experiment C)

Objective: To obtain quantitative data on the electrochemical response of the material tested under sliding at a potential corresponding to the mixed potential recorded in Step B.3.

Tests to be performed:

- *Step C.1*: At the end of Step B.5, immediately start the polarisation of the test sample at a potential E^s_{oc} determined in Step B.3. Maintain this polarisation during the whole of Experiment C.

- *Step C.2*: Start a new sliding test 1 min after polarising the sample at E^s_{oc}. After six contact events, start the electrochemical impedance spectroscopy (EIS) measurements under sliding.
- *Step C.3*: The EIS measurements at E^s_{oc} are started at a frequency of 10^5 Hz and subsequently lowered to at least 10^{-2} Hz by performing measurements at 10 frequencies per decade. The EIS are performed at an ac amplitude of 10 mV. Record the impedance response at each frequency.
- *Step C.4*: On stopping the sliding, immediately remove the test sample from the electrolyte, without delay clean it ultrasonically in ethanol for 5 min, and dry it thoroughly with a dry, cold air dryer. Keep the sample in a desiccator until the start of Experiment D.

Report the EIS data measured during sliding obtained in Step C.3.

7.2.4   Determination of the sliding track profile obtained after testing in the electrolyte (Experiment D)

Objective: To quantify the material loss resulting from Experiments A to C.
   Test to be performed:

- *Step D.1*: After Step C.4, the projected area of the sliding track and the sliding track volume are measured. The area of the sliding track, A_{tr}, is calculated from profilometric, light optical microscopic or SEM measurements. The cross-section of the wear track can be approximated to a semi-ellipse for which the perimeter and area can be easily calculated. This procedure is recommended when the bottom of the wear profile has many irregularities (succession of peaks and valleys). Use for the calculation of the maximum depth measured on each cross section. From these data, the average width e and length L of the sliding track are calculated. The area of the sliding track, A_{tr}, is then:

$$A_{tr} = e\, L \qquad\qquad [7.1]$$

In the case of *rotating sliding tests*, the mass loss, W_{tr}, is calculated from, for example, the average cross-sectional area, S, of the sliding track, its length, L, and the material density, d, as:

$$W_{tr} = S\, L\, d \qquad\qquad [7.2]$$

In the case of *reciprocating sliding tests*, the mass loss, W_{tr} is obtained from either measurements by white light interferometry of the volume of material removed from the surface in the sliding track, and/or by reporting the maximum length and cross-sectional area of the sliding scar obtained, for example, by profilometric measurements.

Report the mass loss, W_{tr}, or the size of the sliding track obtained.

7.3   Tribocorrosion testing procedure under sliding conditions where the material in the sliding track repassivates partially in between successive contact events

This testing is divided into four experiments that are performed on the same test sample which is polished and cleaned before starting the first experiment (see Section

7.1). In order to obtain information on the repeatability of the experiments, it is strongly recommended to perform at least three repeat tests of each experiment.

The sliding tests of interest are performed at open circuit potential, where the counter body moves on the tested flat sample for one cycle with a fixed duration, t_{rot} as defined earlier in this text for rotating ball-on-disc tests, and then remains immobile for a given duration, t_{off}. In the case of reciprocating tests, t_{rot} corresponds to half the sliding frequency. The latency time, t_{lat}, is defined as the time between two successive contacts at a given point in the sliding track, so that:

$$t_{lat} = t_{rot} + t_{off} \qquad [7.3]$$

In general, two values of t_{lat} are selected for these sliding tests. They are taken as $t_{lat\,2} = t_{reac}/1000$, and $t_{lat\,3} = t_{reac}/100$.

By introducing an *off* time during sliding tests, the sliding track can be assumed to consist of two distinct zones, namely:

- a fraction of the sliding track from which the initial passive film has been removed during sliding. This area is referred to as the active area, A_{act}, and
- the remaining sliding track area is covered by a surface film just as before the start of the sliding test. This area is referred to as the repassivated area, A_{repass}.

The fraction of the sliding track surface covered by the passive film, A_{repass}/A_{tr}, is assumed to be proportional to the ratio t_{lat}/t_{reac}. As a result of that proportionality, the relationships between the repassivated area and the total wear track area are at latency times of $t_{lat\,2} = 0.001\,t_{reac}$ and $t_{lat\,3} = 0.01\,t_{reac}$, respectively, $A_{repass\,2} = 0.001A_{tr}$ and $A_{repass\,3} = 0.01\,A_{tr}$, and thus $A_{act\,2} = 0.999\,A_{tr}$ and $A_{act\,3} = 0.99\,A_{tr}$.

To compare the effect of the latency time, t_{lat}, on the different components appearing in the tribocorrosion protocol, the total number of cycles used during these sliding tests must be identical to the total number of cycles performed during the sliding tests done under Section 7.2.

7.3.1 Evolution of the open circuit potential E_{oc} in the electrolyte without any sliding (Experiment E)

Objective: To gather information on the surface characteristics of the passive base material in the absence of any sliding contact.
Test to be performed:

- *Step E.1*: Immerse the sample in the electrolyte and immediately start measuring the open circuit potential of the test sample (E_{oc}) until a stable E_{oc} value is obtained. This is when the long-term fluctuations of E_{oc} are below 1 mV min^{-1} for a minimum of 1 h.

Report the stable E_{oc} value recorded at the end of the experiment.

7.3.2 Evolution of the open circuit potential E^{s}_{oc} in the electrolyte during sliding tests performed at selected latency times (Experiment F)

Objective: To obtain information on the passive–active surface conditions induced by sliding tests performed at different latency times.
Test to be performed: Rotating or reciprocating sliding tests are started immediately after the end of Experiment E.1 without removing the test sample from the electrolyte:

- *Step F.1*: Proceed with measuring E_{oc} of the test sample for 2 min, and record it.
- *Step F.2*: Then load the counter body on the test sample while recording E_{oc} afterwards.
- *Step F.3*: Start the sliding test performed at the selected values of t_{lat}, t_{rot}, and t_{off}. During sliding, the coefficient of friction, μ, can also be recorded. Record on-line the open circuit potential during the whole test.
- *Step F.4*: Stop the sliding test exactly at the time the selected number of contact events is reached, but do not remove the test sample from the electrolyte.
- *Step F.5*: Record the open circuit potential, E_{oc}, during a 10 min period after the end of Step F.4. Then remove the sample from the electrolyte, clean the sample immediately ultrasonically in ethanol for 5 min, and thoroughly dry the sample with a dry, cold air dryer. Keep the sample in a desiccator until the start of Experiment G.

Report the E_{oc} value measured just before starting the sliding tests, the maximum and minimum values of E^s_{oc} measured during sliding, and the final value of E_{oc} measured 10 min after the end of sliding.

Report the coefficient of friction as a function of the sliding cycles or contact events.

7.3.3 Determination of the sliding track profile obtained after tribocorrosion tests (Experiment G)

Objective: To quantify the material loss resulting from Experiments E and F.
   Test to be performed:

- *Step G.1*: The area of the sliding track and the wear track volume are determined on the sample tested under Step F.6. The area of the sliding track, A_{tr}, is calculated in the case of *rotating ball-on-disc sliding tests* from profilometric, optical microscopy or SEM measurements done at several locations uniformly distributed along the wear track using Equation 7.1. The mass loss, W_{tr}, is calculated from Equation 7.2.
   In the case of *reciprocating ball-on-disc sliding tests*, the mass loss, W_{tr} is obtained from either measurements by white light interferometry of the volume of material removed from the surface in the sliding track, and/or by reporting on the maximum length and cross-sectional area of the sliding scar obtained, for example, by profilometric measurements.

Report the mass loss, W_{tr}, or the size of the sliding track obtained.

7.4   Methodology to extract and manage relevant data

Once all of the test data have been collected, these experimental data must be treated to quantify the material degradation in terms of separated mechanical, electrochemical, and synergetic contributions based on the background developed in Chapter 6. The major steps in that data analysis are:

- the management of the electrochemical impedance spectroscopy (EIS) data obtained at open circuit potential (E_{oc}) without any sliding (cf. Experiment A)
- the management of the electrochemical impedance spectroscopy data obtained under sliding at a fixed potential corresponding to E^s_{oc} (cf. Experiment C), and
- the management of the wear data (cf. Experiments D and G).

The analysis of the EIS data represented as Nyquist plots allows us to derive quantitatively the polarisation resistance R_p. In the simplest case, the impedance of the electrode can be represented by an equivalent electrical circuit consisting of a parallel constant phase element (CPE)//R_p circuit in series with the solution resistance R_s:

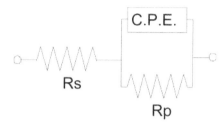

In the Nyquist plane, the impedance plot is an arc of a circle, and R_s and R_p are obtained by extrapolating the arc to the real axis:

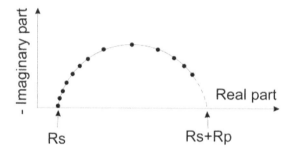

If the impedance plot reveals more than one arc, as for example, in the cases shown hereafter:

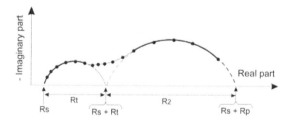

or:

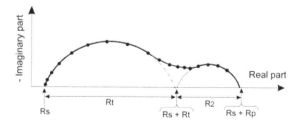

it is necessary to fit the experimental plot with an equivalent circuit model comprising two relaxation time constants, with the following structure for example:

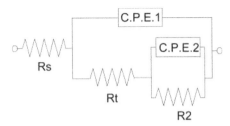

in which R_t is the transfer resistance, and R_p the polarisation resistance ($R_p = R_t + R_2$). When the impedance plots comprise a single arc of a circle, then $R_t = R_p$. When the impedance plots comprise two arcs (indicative of two relaxation time constants), R_p must be derived from the values of R_t and R_2 obtained by fitting the experimental impedance plot using an equivalent circuit model like the above-mentioned circuit. Most electrochemical impedance measurement software provide simulation and fitting facilities with various equivalent circuit models.

A very convenient way to consistently report the test data is shown in Table 7.1. This sheet gives an overview of the key testing parameters and the tribocorrosion test outcome. Such a data reporting is most valuable since it allows a good traceability of the results, for example, in the case of inter-laboratory studies, and a correct distribution of information allowing future use and comparison of everybody's experimental results.

An illustration of management of the data generated by the tribocorrosion protocol designed in Chapter 6, is given based on a case study presented in Section 7.5.

7.5   Case study: Quantitative calculation of tribocorrosion components for AISI 316L immersed in a corrosive solution

In this section, the tribocorrosion protocol is applied to a model system with the objective to illustrate its usefulness for the quantitative determination of the different components contributing to material loss in a material system subjected to tribocorrosion [1]. Electrochemical and sliding tests were performed under tribocorrosion conditions on a passivating material, namely AISI 316 stainless steel, immersed in a dilute sulphuric acid solution. Surface characteristics and amount of material loss are affected by the combined periodic mechanical removal of the passive surface film on AISI 316L during sliding, and its subsequent electrochemical re-growth between two successive contact events. This phenomenon is analysed by varying the latency time, and by determining its influence on the wear volume.

7.5.1   Experimental testing conditions used to investigate the tribocorrosion components for AISI 316L tested in dilute sulphuric acid [2]

AISI 316 stainless steel cylindrical samples with a diameter of 25 mm and a height of about 20 mm were tested in 0.5 M H_2SO_4. Before starting the tests, these samples were polished with 1 μm diamond paste followed by ultrasonic cleaning in acetone for 5 min and rinsing in ethanol also for 5 min. The flat circular top part of the cylinders with an area, A_o, of 4.91 cm^2 was exposed to the electrolyte. All electrochemical and rotating ball-on-disc sliding tests were performed at 25°C. A platinised titanium gauze was used as the counter-electrode, and a mercury/mercury sulphate/saturated

Table 7.1   Sheet of key parameters and test results of tribocorrosion test to improve procedures and traceability of results

---

GUIDELINES FOR THE INTER LABORATORY STUDY

Tribocorrosion protocol for passivating materials

Name of test laboratory:

Experimental Arrangement

Contact configuration: *ball loaded on top of flat sample (disc)*

Trade Mark potentiostat:

Ref. Elect. used: (*vs.* SHE)

Area of Pt-counter electrode in contact with electrolyte =   (cm^2)

Area (A_0) of test sample in contact with electrolyte =   (cm^2)

Distance between Ref. Elect. and rim of wear track =   (mm)

Volume of electrolyte* in each experiment =   (mL)

Standard volume 50 mL/cm^2 of sample area in contact with the electrolyte

Solution pH: Other:

Counter body radius: alumina ball, diameter =   (mm)   Other: (mm)

Normal load: (N)   Other: (N)

Value of t_{reac}:   (s) E_{oc} =   (V vs. Ref. Elect.) E^s_{oc} =   (V vs. Ref. Elect.)

Continuous sliding: (YES) or (NO)

Standard rotation period of ball around disc axis t_{rot} =   (s)

Reciprocating sliding: (YES) or (NO)

Standard sliding frequency =   (Hz)

Standard peak-to-peak displacement amplitude =   (mm)

Total number of events*:

* *one event is one mechanical contact at a given point*

Period between two successive contacts at a given point: t_{lat1} =   (s)

                                                                                                t_{lat2} =   (s)

Results

Size of wear track: length =   (μm)   width* =   (μm)   depth* =   (μm)

* *In the middle of the wear track*

Total distance sliding: (m)

Total duration of the tribocorrosion test: (s)

Type of equivalent electrical circuit used:

r_{pass} =	($\Omega\ cm^2$)		i_{pass} =	(mA cm^{-2})
r_{act} =	($\Omega\ cm^2$)		i_{act} =	(mA cm^{-2})

W_{tr}=   (cm^3) W^c_{repass}=   (cm^3) W^c_{act}=   (cm^3) W^m_{repass}=   (cm^3)

---

potassium sulphate electrode (SSE) as the reference electrode. For EIS, a sinusoidal potential variation of 10 mV$_{RMS}$ was superimposed at frequencies of 10 kHz down to 10 mHz, on the E_{oc} of the sample recorded before or during sliding tests. Counter bodies made of zirconia pins with a spherical tip radius of 100 mm, were loaded on top of the cylindrical test samples at a normal load of 5 N, corresponding to a maximum Hertzian contact pressure of 140 MPa. The counter body was positioned 5 mm eccentric from the rotation axis, corresponding approximately to the half radius of the test samples.

7.5.2   Results obtained from the experiments described in Sections 7.2 and 7.3, and their analysis

The evolution of the open circuit potential with time starting from immersion is shown in Figure 7.1. It provides information on the electrochemical reactivity of the surface of the test sample. In the case of AISI 316, the open circuit potential, E_{oc}, is initially around –0.9 V *vs.* SSE, and increases by 150 mV during the first hour of immersion. Then, a rapid increase in E_{oc} is observed (>2.5 V h^{-1}) followed by a stabilisation at around –0.3 V *vs.* SSE (<60 mV h^{-1}). Based on these experimental data, the surface state of the sample may be considered as stabilised within 5000 s of immersion. This period of 5000 s sets the value of t_{reac}. In this electrolyte, the equilibrium potential of the reduction reaction of hydrogen ions is around –0.7 V *vs.* SSE. The open circuit potential recorded indicates that the reduction reaction taking place at the surface of AISI 316 is most probably the reduction of hydrogen upon immersion, but becomes the reduction of oxygen dissolved in the electrolyte after 4000 s.

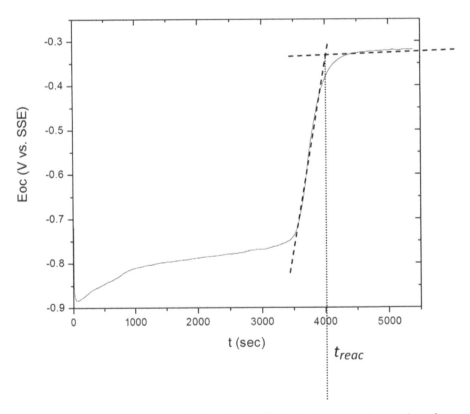

7.1 Evolution of the open circuit potential (E_{oc}) with time upon immersion of a freshly polished AISI 316 sample in 0.5 M H$_2$SO$_4$ at 25°C in the absence of any mechanical loading

The surface state of AISI 316 stainless steel was then investigated by electrochemical impedance spectroscopy performed when a stable open circuit potential was reached. Such a measurement is shown in Figure 7.2 as a Nyquist plot. This shows that the imaginary part of the impedance, Z_{im}, versus the real part, Z_{re}, is an arc of a circle. Such an impedance plot corresponds to an equivalent electrical circuit in which the solution resistance, R_s, is in series with a circuit consisting of a constant phase element, C.P.E., in parallel with a polarisation resistance, R_p. The values of R_s and R_p are obtained by extrapolating the arc to the real axis. The specific polarisation resistance (value per unit of area), r_{pass}, and the corrosion current density, i_{pass}, are calculated from Equations 6.5 and 6.6, respectively, developed in Chapter 6. The values of r_{pass} and i_{pass} reported in Table 7.2 indicate that AISI 316 stainless steel can be considered at its stable open circuit potential without sliding and covered by a passive surface film. The current density i_{pass} can thus be linked to the dissolution of the stainless steel through its surface film which is in a stationary passive state at the open circuit potential.

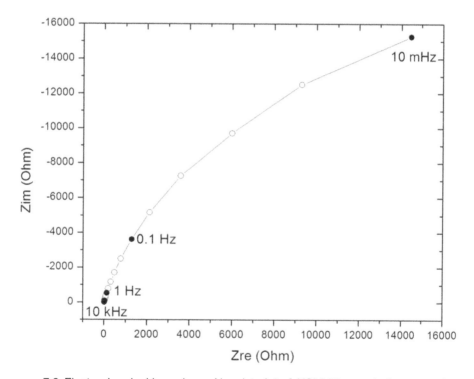

7.2 Electrochemical impedance Nyquist plot of AISI 316 recorded at open circuit potential, after stabilisation in 0.5 M H$_2$SO$_4$ at 25°C

Table 7.2 Specific polarisation resistance and corrosion current density of AISI 316 immersed in 0.5 M H$_2$SO$_4$ and measured at its stable open circuit potential at 25°C under mechanically unloaded conditions

A_o (cm^2)	R_p (Ω)	r_{pass} (Ω cm^2)	i_{pass} (A cm^{-2})
4.91	42.56×10^3	209×10^3	1.148×10^{-7}

The evolution of the open circuit potential with immersion time (Figure 7.1) and the specific impedance values (Table 7.2) indicate that a stable passive surface state is reached on AISI 316 stainless steel immersed in 0.5 M sulphuric acid after 5000 s. The value t_{reac} is then used to calculate the rotation period t_{rot}, and the latency times t_{lat}. The calculated value of $t_{rot} = t_{reac}/10\ 000$ is then 0.5 s which corresponds to a rotation rate of the counter body of 120 rpm.

As the next step, rotating ball-on-disc sliding tests were started at the time the open circuit potential stabilised after immersion. The data obtained on AISI 316 stainless steel immersed in 0.5 M H_2SO_4 are shown in Figure 7.3. Before the initiation of sliding, the formation of a passive surface film increases the open circuit potential, up to a value of about –0.3 V *vs.* SSE, as previously observed in Figure 7.1. At the start of the sliding, the open circuit potential drops sharply to more negative potential values revealing a modification of the surface state in the sliding track area on the test material. The sliding track area, A_{tr}, is for the sample size selected in this study, however only about 10% of the sample area, A_o, is exposed to the electrolyte. The E_{oc} value recorded under sliding is thus a mixed potential resulting from the coupling of a material with different surface states inside and outside the sliding track. During rotating ball-on-disc sliding tests, the open circuit potential remains more or less constant at that lower value. At the time that sliding is ended, the open circuit potential starts to rise in a rather exponential way. This indicates a progressive evolution of the surface state of the material in the sliding track area due to a restoration of a passive surface state condition on it. This is supported by the fact that the open

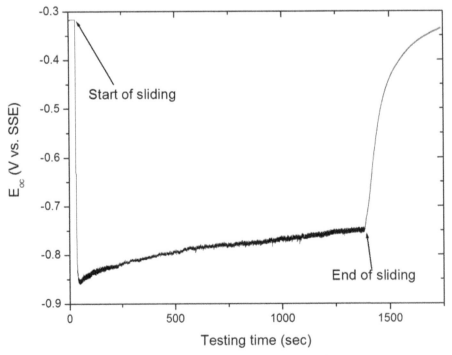

7.3 Evolution of the open circuit potential, before, during, and after continuous unidirectional sliding tests performed at 5 N and 120 rpm on AISI 316 immersed in 0.5 M H_2SO_4 at 25°C

circuit potential returns to a value corresponding to that measured before sliding started.

Impedance measurements were carried out during rotating ball-on-disc sliding tests while the sample was polarised at the mean open circuit potential recorded during sliding (−0.8 V *vs.* SSE). The impedance plot obtained is presented in Figure 7.4. From this Nyquist plot, the polarisation resistance during sliding, R_{ps}, was measured. A polarisation resistance of 2500 Ω was obtained. This polarisation resistance is in fact the combination of two polarisation resistances connected in parallel. The first one is related to the sliding track area, A_{tr}, and the second one to the area, A_o-A_{tr}, outside the sliding track. In a first approximation, the whole sliding track area, A_{tr}, can be considered during such sliding conditions to be in an active electrochemical state (corrosion) because the time period between successive contact events ($t_{rot} = 0.5$ s) is small enough compared to t_{reac} ($t_{rot} = t_{reac}/10\ 000$) to prevent the restoration of the passive film at any place in the sliding track area. The specific polarisation resistance for the active material, r_{act}, and the current density, i_{act}, were calculated as explained and developed in Chapter 6 (Section 6.2.3) and are given in Table 7.3.

The comparison of the specific polarisation resistances given in Tables 7.2 and 7.3 indicates that AISI 316 stainless steel immersed in 0.5 M sulphuric acid at E_{oc} until full stabilisation of its surface state without sliding, is covered with a passive surface film. Under sliding, there is a significant degradation of that protective surface film in the sliding track.

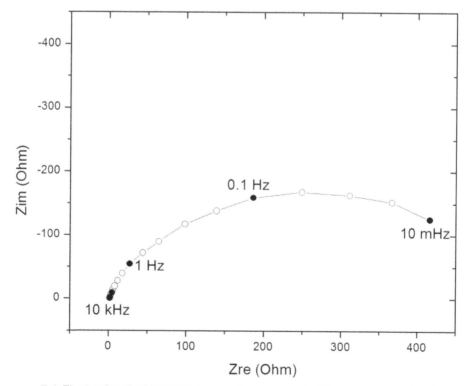

7.4 Electrochemical impedance spectra measured at the mean open circuit potential value during continuous unidirectional sliding tests performed at 5 N and 120 rpm on AISI 316 immersed in 0.5 M H_2SO_4 at 25°C

Table 7.3 Sliding track area, specific polarisation resistance, and corrosion current density, of AISI 316 in 0.5 M H_2SO_4, at open circuit potential under continuous unidirectional sliding at 5 N, 120 rpm, and 25°C

A_{tr} (cm²)	r_{act} (Ω cm²)	i_{act} (A cm⁻²)
0.164	84	2.8×10^{-4}

The properties of surface films formed on AISI 316 stainless steels immersed in 0.5 M sulphuric acid were investigated by rotating ball-on-disc sliding tests performed at different latency times. Such tests consist of sequences of one sliding cycle (duration $t_{rot} = 0.5$ s) followed by a pause for a given time t_{off}. These tests allow the analysis of the periodic removal and re-growth of surface films reflected in a cyclic evolution of the open circuit potential. The stop time, t_{off}, is introduced after each sliding cycle to allow the bare material in the sliding track to react with the surrounding medium resulting in a partial or full re-growth of a surface film. This is thus a self-healing process. The latency time is selected so that the re-growth of a surface film in between two successive contact events is not negligible as was the case in the rotating sliding tests performed at low latency times generating active materials in the sliding tracks. In these tests, t_{off} was set at 4.5 s and 49.5 s, so that $t_{lat\,2} = t_{reac}/1000 = 5$ s, and $t_{lat\,3} = t_{reac}/100 = 50$ s, in which t_{reac} is the reactivity time necessary for a full stabilisation of the surface film.

The evolution of the open circuit potential with time during rotating ball-on-disc sliding tests on AISI 316 stainless steel immersed in 0.5 M H_2SO_4 is shown in Figure 7.5. It illustrates the effect of the off-time on the outcome of these tests.

During rotating ball-on-disc sliding tests with an off-time of 0 s, the open circuit potential is stable around –0.82 V *vs.* SSE. During rotating ball-on-disc sliding tests performed at an off-time of 4.5 s, the open circuit potential drops during the on periods and rises during the off periods. This results in fluctuations of the open circuit potential between –0.82 and –0.74 V *vs.* SSE. A similar trend is noticed at an off-time of 49.5 s but with larger fluctuations of the open circuit potential between –0.73 and –0.59 V *vs.* SSE. These fluctuations of the open circuit potential reveal the ability of AISI 316 immersed in 0.5 M sulphuric acid to recover its surface film in the sliding track in between successive sliding contacts. This self-healing effect increases with increasing latency time. At $t_{off} = 0$ s and $t_{off} = 4.5$ s, the same value of the open circuit potential during sliding is obtained. This indicates that a t_{off} of 4.5 s is too small to allow a significant restoration of the passive film in the wear track during the off period. The marked increase in the potential during sliding at $t_{off} = 49.5$ s, indicates that the passive film is significantly restored in the sliding track during the off periods, and that the passive film in the wear track is only partially destroyed during the sliding period. The increase in the potential at the end of the off period, as well as the increase in the amplitude of the potential variation with t_{off}, indicate that partial restoration of the passive film in the sliding track during the off-time, takes place even at $t_{off} = 4.5$ s. The growth of the passive film in the sliding track will even take place at $t_{off} > 49.5$ s because the maximum value of the potential noticed at $t_{off} = 49.5$ s is still below the stabilised open circuit potential value recorded before sliding.

In order to calculate the corrosive and mechanical contributions to the material loss as described in Chapter 6, Section 6.3, weight content, molecular weight, number of exchanged electrons, and density of the tested material are all required. These parameters are summarised in Table 7.4.

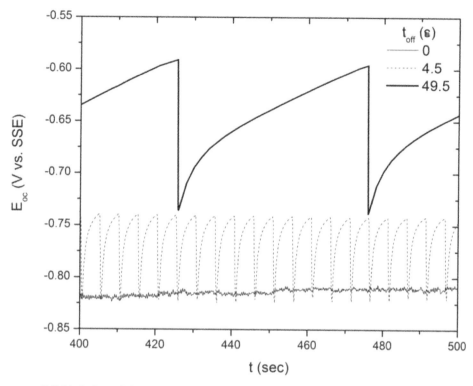

7.5 Variation of the open circuit potential during continuous (t_{off} = 0 s) and intermittent (t_{off} = 4.5 s or 49.5 s) unidirectional sliding tests with AISI 316 against zirconia pins immersed in 0.5 M H_2SO_4 at a normal force of 5 N

Table 7.4   Weight content, number of exchanged electrons and molecular weight of constituents of AISI 316, corresponding value of C (Chapter 6, Equation 6.23) and density of the alloy

x_{Fe}	n_{Fe}	M_{Fe} (g mol^{-1})	x_{Ni}	n_{Ni}	M_{Ni} (g mol^{-1})	x_{Cr}	n_{Cr}	M_{Cr} (g mol^{-1})	C (g mol^{-1})	d (g cm^{-3})
0.72	2	56	0.1	2	59	0.18	3	52	25.3	7.9

As described in Chapter 6, it is now possible to calculate the material loss due to corrosive wear of the active material, W_{act}^c from Equation 6.10. The material loss due to mechanical wear of the active material W_{act}^m can be calculated from Equation 6.11, and the volume of the sliding track measured at the end of the test, W_{tr}. In the same manner, the material loss of the repassivated material due to corrosive and mechanical wear under sliding conditions with two different off-times can be calculated using Equations 6.13 and 6.14. The results obtained by the procedure detailed above, are summarised in Table 7.5. All wear values are expressed in volumetric material loss per cycle, obtained by dividing the mass losses by the density *d* and the number of contact events, *N*. The wear results of Table 7.5 are plotted in Figure 7.6.

Table 7.5 Experimental test outcome and calculated tribocorrosion components obtained on AISI 316 tested under continuous (t_{lat} = 0.5 s) and intermittent sliding (t_{lat} = 5 and 50 s) in 0.5 M H_2SO_4. Unidirectional sliding was carried out at 5 N and 120 rpm

t_{lat} (s)	A_{trmin} (cm²)	A_{trmax} (cm²)	A_{act} (cm²)	A_{repass} (cm²)	W_{tr} (cm³/cyc.)	W^c_{act} (cm³/cyc.)	W^m_{act} (cm³/cyc.)	W^c_{repass} (cm³/cyc.)	W^m_{repass} (cm³/cyc.)
0.5	0.081	0.164	0.123	0	4.75×10^{-9}	5.83×10^{-10}	4.17×10^{-9}	0	0
5	0.081	0.327	0.204	2.0×10^{-4}	1.75×10^{-8}	2.92×10^{-9}	6.93×10^{-9}	3.90×10^{-15}	7.63×10^{-9}
50	0.081	0.416	0.246	2.5×10^{-3}	2.87×10^{-8}	1.46×10^{-8}	8.36×10^{-9}	4.75×10^{-13}	5.68×10^{-9}

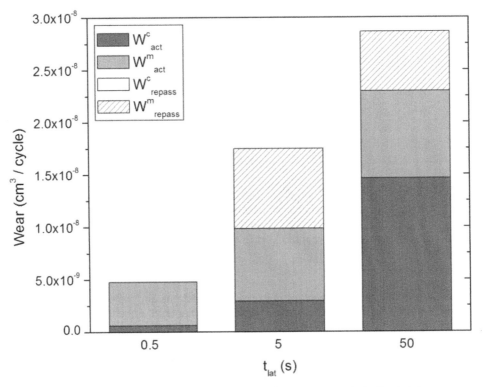

7.6 Contribution of the different tribocorrosion components to the total volumetric material loss in the sliding track on AISI 316 immersed in 0.5 M H_2SO_4, for unidirectional sliding tests performed at different latency times, t_{lat}

For the interpretation of the outcome of tribocorrosion tests, as described in Chapter 6, the specific material loss components in the sliding track, W^j_n, are calculated per unit area and per cycle, according to Equation 6.31. Additionally, the ratios K_c and K_m are calculated based on Equations 6.29 and 6.30, respectively. The results are shown in Table 7.6.

The value K_c obtained under sliding conditions with a latency time equal to 0.5 s indicates that the contribution of mechanical wear is prevailing, in spite of the low value of the contact pressure and the high corrosiveness of the 0.5 M H_2SO_4 solution. At higher t_{lat} values, namely at t_{lat} = 5 and 50 s, the contribution of the corrosive wear

Table 7.6   Calculated specific wear components, and K_c and K_m ratios

t_{lat} (s)	W^c_{act} (cm/cycle)	W^m_{act} (cm/cycle)	W^c_{repass} (cm/cycle)	W^m_{repass} (cm/cycle)	K_c	K_m
0.5	4.75×10^{-9}	3.39×10^{-8}			1.39×10^{-1}	
5	1.43×10^{-8}	3.39×10^{-8}	1.91×10^{-11}	3.73×10^{-5}	2.01×10^{-1}	9.10×10^{-4}
50	5.94×10^{-8}	3.39×10^{-8}	1.91×10^{-11}	2.28×10^{-6}	1.04	1.48×10^{-2}

increases, and becomes of the same order as the contribution of mechanical wear at $t_{lat} = 50$ s.

The small K_m ($\ll 1$) indicates that the passive film is much more sensitive to mechanical wear than the active bare material. The increase in K_m with t_{lat} is explained by the increased mechanical wear resistance of the passive film when it becomes thicker. However, even at $t_{lat} = 50$ s, the resistance of the passive film to mechanical wear is much lower than the resistance to mechanical wear of the bare material.

7.6   Conclusions

The tribocorrosion test protocol for passivating materials was validated for the investigation of the tribocorrosion behaviour of AISI 316 stainless steel in 0.5 M H_2SO_4 by rotating ball-on-disc sliding tests performed at different latency times. Under certain sliding conditions, monitoring of the surface state of the sample by electrochemical methods revealed the removal and re-growth of the passive surface film. An analysis was carried out on the evolution of the passive film in the wear track as well as its effect on the components of the corrosive or mechanical degradation acting during tribocorrosion. The material loss depends on the thickness of the passive surface film indicating that there is a strong interaction of the tribocorrosion process on the surface characteristics in the sliding track. A quantitative data analysis was also performed based on the specific material losses. It allowed the determination of the respective contributions of mechanical and corrosive loads acting on the active and repassivated parts in the sliding tracks.

Our insight into the mechanical–chemical behaviour of surfaces and, more particularly, the tribocorrosion of passivating metallic materials, will progress greatly thanks to the protocol analysis methodology proposed in this chapter. Its implementation in research centres and in industry opens the way to achieve repeatable and reproducible test data on tribocorrosion, and thus to support further material developments necessary to solve actual industrial problems of material degradation and its related hazardous environmental and health effects.

By helping to improve capitalisation and the sharing of knowledge, this approach is a step towards building up a quality process within European research and development in the sense of ISO 9001.

References

1.   N. Diomidis, J. P. Celis, P. Ponthiaux and F. Wenger: *Lubr. Sci.*, 2009, **21**, (2), 53–67.
2.   N. Diomidis, J. P. Celis, P. Ponthiaux and F. Wenger: *Wear*, 2010, **269**, (1–2), 93–103.

8
Normative approach

Amaia Igartua

TEKNIKER Avda. Otaola 20, P.K. 44, SP-20600 EIBAR Gipuzko, Spain
aigartua@tekniker.es

Mª Pilar Gomez-Tena

Instituto de Tecnología Cerámica, Campus Universitario Riu Sec |
Av. Vicent Sos Baynat s/n | 12006 Castellón, Spain
pilar.gomez@itc.uji.es

8.1   Why a standardisation?

The design, implementation and application of any product or activity involve a large number of people, separated from each other both in time and distance, which are in most of cases not interrelated. This supports the need for common rules which must be followed everywhere. Manufacturers, users and consumers are aware of the product, the activity or the service that they have in their hands. Standards are a tool that allows filling up this gap by introducing a series of guidelines and specifications.

A **standard** is thus a document containing technical specifications based on experience and technological development which regulates a product or a particular activity. Thanks to standards, the control and monitoring of raw materials, machines, articles, processes, trials, services, etc. can be done and repeated anywhere and at any time. The rules apply to any product and field of activity from simple to complex products, and from a sophisticated nuclear reactor down to a steel fork used daily. A rule can be created by any person or group of people who know the product or activity, and it can be applied at the international, national or local level. There are a large number of companies, bodies or entities that have their own internal rules which regulate a number of activities, products or processes. Now, if the rule is external and international, the effectiveness will be higher since everybody will know the product, and may manufacture, control or use the product specifications.

Standardisation is the voluntary process of developing technical specifications based on consensus among all interested industrial parties including small- and medium-sized enterprises (SMEs), consumers, trade unions, environmental non-governmental organisations (NGO), public authorities, etc. It is carried out by independent standard bodies, acting at the national, European, and international level. Nowadays, there are national and international organisations governing the creation and distribution of standards like ISO (International Organization for Standardization), EN (European Standards), ASTM (American Society for Testing and Materials), DIN (Deutsches Institut für Normung), UNE (Unification of Spanish Standards), etc.

Of course, scientific and technological developments as well as their use evolve with time, in turn leading to modifications or alterations of products or activities. For that reason, standards are living documents which require periodic reviews so they are updated based on developments in knowledge and technology. While the use of standards remains voluntary, the European Union has, since the mid-1980s, promoted an increasing use of standards to support its policies and legislation. Standardisation has contributed significantly to the completion of the Internal Market in the context of 'New Approach' legislation which refers to European standards developed by the European standard organisations. Furthermore, European standardisation supports European policies in the areas of competitiveness, information and communication technologies (ICT), innovation, interoperability, environment, transport, energy, consumer protection, etc.

Standardisation is thus an excellent tool to facilitate international trade, competition, and the acceptance and recognition of innovations by markets. A key challenge for European standardisation is to strengthen its contribution to the competitiveness of SMEs. Furthermore, standardisation promotes rationalisation and quality assurance of enterprises and markets in general, because it is an important development tool to broaden actual knowledge, and promotes innovation.

As the area of standardisation gains a higher profile internationally, a number of studies have reviewed its benefits. The findings of a study on the economics of standardisation for the British Department of Trade and Industry include the following [1]:

- Standardisation is a key part of the microeconomic infrastructure; it can enable innovation and act as a barrier to undesirable outcomes.
- Standardisation increases competition and that does not necessarily increase the profitability of companies. However, it is in the interests of the economy as a whole.
- It is clear that traditional public standards setting procedures are under pressure. It is widely perceived that they are not 'fast enough'.
- There is doubt that a producer led standardisation process can give full account to customer interests.

Apart from all these benefits, it must also be mentioned that standardisation improves some technical features of the cycle production, like:

Cycle time: Standardisation reduces cycle time and assists with production smoothing. Standardised processes can also help to eliminate non-value adding time in the organisation, and reduce unnecessary motion.

Stability: Standardisation will increase stability in the organisation with a clear start and stop in processes.

Safety and quality: The unexpected can increase the risk in the organisation and create a safety hazard. Standardised processes can also increase quality standards in the organisation.

Analysis: Standardisation simplifies the auditing process and makes it easier to identify opportunities in the system.

Visual management: Standardised processes will also make it easier to visualise processes with images and photos.

Training: Standardisation makes the training of staff easier and getting new employees 'up to speed'. It takes the guesswork out of the business, and assists the organisation in creating standardised operation procedures (SOPs).

8.2   Technical and economic benefits in the field of ceramic and metallic materials

Standards are reference points to understand the level of quality and safety of products and services, and to allow exchanges at the national and international level. For example, if a company in a country produces a screw following the international manufacturing standard (e.g. ISO 898-1), it may be sold or marketed in any other country since everyone knows the mechanical properties of that screw and thus its quality. Therefore, the use or application of the rules favours the elimination of technical and commercial barriers. Thus, the rules act as a fundamental tool for industrial and commercial development as well as for research and innovation.

In fact, this is reflected in both technical and economic benefits which are very difficult to calculate but which undoubtedly reach billions of Euros. It must be borne in mind that in the absence of rules, each time someone would like to purchase a product or service, specifications and guidelines should be developed that follow the provider control. Likewise, anyone who wants to sell a product or service should validate or prove to his customer that the article is secure and meets their specific requests. This would be time consuming leading to high costs, and would hamper its marketing causing a slowdown in development and innovation. Product marks and certifications are based on the fulfilment of a series of norms or directives, and when consumers buy a product, they fix and mark the corresponding certificate. Indeed, we have pressed far beyond the era where the almost casual use of lubricant was sufficient to alleviate wear problems. In a report from the Hanniker Conference on National Material Policy [2], tribology is listed as one of the three major technical areas affecting material utilisation and cost reductions. The enormous cost of tribological deficiencies to any national economy is mostly caused by the large amount of energy and material losses occurring simultaneously on virtually every mechanical device in operation. When reviewed on the basis of a single machine, the losses are small. However, when the same loss is repeated on perhaps a million similar machines, then the costs become very large.

There are many examples which suggest that a form of 'tribology equation' can be used to obtain a simple estimate of either costs or benefits from existing or improved tribocorrosion practice [3]. Such an equation can be summarised as:

$$\text{Total Tribocorrosion Cost/Saving} = \text{Sum of Individual Machine Costs/Savings} \times \text{Number of Machines}$$

This equation can be applied to any problem to estimate the relevance of tribocorrosion in a particular situation (see Figure 8.1).

In 1966, Sir P. Jost estimated that by applying the basic principles of tribology, the UK economy could save approximately £515 million per annum at 1965 values [4]. A similar report published in West Germany in 1976 revealed that the economic losses caused by friction and wear cost about 10 billion DM per annum, at 1975 values, which is equivalent to 1% of the Gross National Product [5]. About 50% of these losses were due to abrasive wear. In the USA, it has been estimated that about 11% of total annual energy could be saved in the four major areas of transportation, turbo machinery, power generation, and industrial processes through progress in tribology. For example, tribological improvements in cars alone could save about 18.6% of total annual energy consumed by cars in the USA, which is equivalent to about 14.3 billion US$ per annum. In the UK, the possible national energy savings

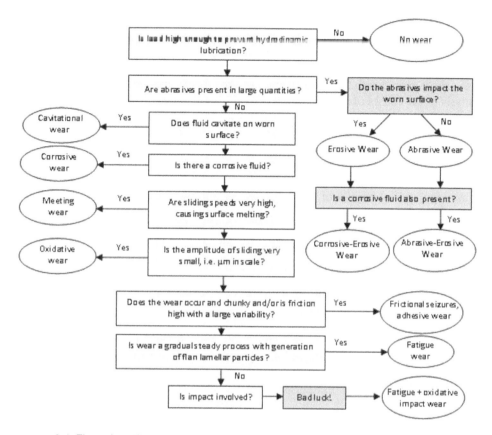

8.1 Flow chart illustrating the relationship between operating conditions and type of wear [3]

achieved by the application of tribological principles and practices have been estimated to be between £468 and £700 million per annum.

The economics of tribology and tribocorrosion are of such gigantic proportions that tribological programmes have been established by industry and governments in many countries throughout the world. The economic losses caused by corrosion have been quantified in 26 industrial sectors in the USA. The estimated direct cost was 276M$/year in 1998, which represented 3.1% of the Gross National Product. The production of steels and the improvement in their mechanical properties make this material very useful, but it is very prone to corrosion, since 25% of the annual production of steel is destroyed by corrosion. Corrosion can also affect security. In 1984, in Mexico, the explosion of a GLP gas container, due to exhaust gases produced by corrosion, caused 400 deaths. The cost of repairing or replacing water main piping perforated by corrosion is becoming a major item in many municipal budgets, and correspondingly in most individuals' municipal taxes. In metropolitan Toronto, for instance, it is estimated that over $5 million will be spent in 1983 for repair, replacement, or renewal of domestic water piping systems. Winnipeg, which has perhaps the highest failure rate of any Canadian city, had approximately 2200 failures in 1982, requiring a $7.7 million programme just to limit annual leaks to the 2200 level, which is a leak frequency of $1.1 \text{ km}^{-1} \text{ year}^{-1}$.

Tribocorrosion phenomena are encountered in many technological areas where they cause damage to installations, machines and devices. Often, tribocorrosion damage is a problem for safety or for human health.

Material losses owing to friction and wear can potentially be avoided or diminished, provided that proper tribological practice is applied where required. The analysis of the causes of friction and wear, and their standardisation can have direct commercial implications, even in terms of who bears the cost of excessive wear or friction.

As soon as the extent of economic losses due to friction and wear became clear, researchers and engineers rejected many of the traditional limitations to mechanical performance. Some technological improvements are so radical that the whole technology and economics of the development impose a removal of oil and the use of a dry lubricating system such as high-temperature-resistant and self-lubricating materials. If the engine can operate adiabatically at high temperatures, heat previously removed by the now obsolete radiator can be turned to mechanical work. As a result, a fuel efficient, lightweight engine might be built which will lead to considerable savings in fuels, oils and vehicle production costs. A fuel efficient engine is vital in reducing transportation and agricultural costs and therefore is a very important research and development task.

Other examples of such innovations include surface treated cutters for sheep shearing, surface hardened soil engaging tools, polyethylene pipes for coal slurries and ion implanted titanium alloys for orthopaedic endoprostheses. Whenever wear and friction limit the function or durability of a device or appliance, there is scope for tribology and tribocorrosion to offer some improvements.

8.3   Standardisation process

International standards are developed by technical committees and subcommittees by a process divided into six steps [6, 7]:

- Step 1: *Proposal stage*
- Step 2: *Preparatory stage*
- Step 3: *Committee stage*
- Step 4: *Enquiry stage*
- Step 5: *Approval stage*
- Step 6: *Publication stage*.

The technical committees (TC) and subcommittees (SC) may set up working groups (WG) of experts for the preparation of a Working Draft. Subcommittees may have several working groups, which can have several sub groups (SG) [8]. Table 8.1 gives a full overview of the stages in the development of an ISO standard [9–12].

The first step is a proposal of work (New Proposal) that has to be approved at the relevant subcommittee or technical committee (e.g. SC29 and JTC1, respectively in the case of Moving Picture Experts Group (MPEG) – ISO/IEC JTC1/SC29/WG11). A working group (WG) of experts is then set up by the TC/SC for the preparation of a Working Draft. When the scope of a new work is sufficiently clarified, some of the working groups (e.g. MPEG) usually make an open request for proposals known as a 'Call for proposals'. The first document that is produced, for example, for audio and video coding standards is called a Verification Model (VM) (previously also called a Simulation and Test Model). When sufficient confidence in the stability of

Table 8.1   Stages in the development process of an ISO standard [6–9]

Stage code	Stage	Associated document name	Abbreviations
00	*Preliminary stage*	Preliminary work item	PWI
10	*Proposal stage*	New work item proposal	NP or NWIP, NP Amd/TR/TS/IWA
20	*Preparatory stage*	Working draft(s)	AWI, AWI Amd/TR/TS, WD, WD Amd/TR/TS
30	*Committee stage*	Committee draft(s)	CD, CD Amd/Cor/TR/TS, PDAmd (PDAM), PDTR, PDTS
40	*Enquiry stage*	Enquiry draft	DIS, FCD, FPDAmd, DAmd (DAM), FPDISP, DTR, DTS
50	*Approval stage*	Final draft International Standard	FDIS, FDAmd (FDAM), PRF, PRF Amd/TTA/TR/TS/Suppl, FDTR
60	*Publication stage*	International Standard	ISO TR, TS, IWA, Amd, Cor
90	*Review stage*		ISO TR, TS, IWA, Amd, Cor
95	*Withdrawal stage*		

the standard under development is reached, a Working Draft (WD) is produced. This draft is in the form of a standard but is kept internal to the working group for revision. When a Working Draft is sufficiently solid and the working group is satisfied that it has developed the best technical solution to the problem being addressed, it becomes a Committee Draft (CD). If required, it is then sent to the P-members of the TC/SC (National Bodies) for ballot.

The Committee Draft becomes a Final Committee Draft (FCD) if the number of positive votes is above the quorum. Successive committee drafts may be considered until consensus is reached on the technical content. When this is reached, the text is finalised for submission as a draft International Standard (DIS). The text is then submitted to National Bodies for voting and comments within a period of 5 months. It is approved for submission as a final draft International Standard (FDIS) if a two-thirds majority of the P-members of the TC/SC are in favour and not more than one-quarter of the total number of votes cast are negative. ISO will then hold a ballot with National Bodies where no technical changes are allowed (yes/no ballot), within a period of 2 months. It is approved as an International Standard (IS) if a two-thirds majority of the P-members of the TC/SC is in favour and not more than one-quarter of the total number of votes cast are negative. After approval, only minor editorial changes are introduced into the final text. The final text is sent to the competent standardisation centre which publishes it as a Standard.

During the preparatory stage, proficiency testing by inter-laboratory comparison is generally required (ISO/IECC 43-1:1997) [13, 14].

8.4   State-of-the-art on tribocorrosion normative scenarios

One of the major difficulties in tribocorrosion studies is the inter-laboratory comparability of results and findings because of a lack of a standard test apparatus.

- **Proper insulation**. In a tribocorrosion test system, the sample is subjected to an electric current therefore, it is very important to eliminate any leakage of the current through the contact or part of the circuit of the conducting materials in the system.

- **Counter body**. If the two contacting surfaces are conductive, the collection of responses related to corrosion can be complicated. Hence, generally, one of the contacting bodies is kept as an insulator and the conductive body is studied for its tribocorrosion resistance.
- **Dynamic tribocorrosion test system and understanding the synergism**. A corrosion test favours a stabilised system for starting the tests, and studying changes in surface chemistry under the influence of solution or environment.

8.4.1   Standards of interest to industrial fields

Standards of interest for ceramics

Ceramics are generally defined as inorganic non-metallic solid materials. This definition includes not only materials such as traditional ceramics, for example, pottery, porcelains, refractories, cements, abrasives and glass, but also non-metallic magnetic materials, ferroelectrics and a variety of other new materials.

For traditional ceramics, there is no information available on their tribocorrosion behaviour, since this kind of ceramic is highly resistant to the products that they are in contact with in their daily life. Corrosive processes such as those considered in tribocorrosion do not degrade these traditional ceramics.

However, a renewed interest in ceramics is rooted in unique materials classified as advanced structural ceramics, electronic and optical ceramics. Structural ceramics are at present used in diverse applications as tribomaterials due to their unique properties that include resistance to wear and corrosion at elevated temperatures, low density and unique electrical, thermal and magnetic properties. Applications include precision instruments such as bearings, cutting tool inserts, prosthetic articulating joints, and engine components.

Over the past 20 years, a large number of studies have been carried out on the tribological properties of structural ceramics. Nevertheless, there is no specific standardised normative for this kind of material, although some authors have documented different tribocorrosion methodologies in their own research programmes, above all applied to ceramic structural materials [15]. Among them, it is possible to emphasise different tests carried out on ceramics such as alumina [15], silicon nitride [16], silicon carbide [17], zirconia [18], titanium or chromium nitride [19, 20], tungsten carbide [16] and other composites [21, 22].

Standards of interest for metals

The study of the synergistic effect between wear and corrosion on metallic surfaces is quite complex and involves knowledge of the tribological and electrochemical aspects of the mechanical system under study. For developing tribocorrosion experiments, it is necessary to control not only the mechanical but also the chemical test conditions. Several chemical and mechanical parameters do indeed affect the system response. Tribological parameters can be the applied load, the sliding speed, the friction coefficient, etc. Electrochemical parameters are the electrolyte (pH, concentration of aggressive species), the potential of the electrode, etc. In aqueous ionic electrolytes, electrochemical techniques offer the possibility to control *in situ* the surface modification of metals and coatings. The electrochemical techniques allow us to correlate chemical reactions with the mechanical behaviour of the tribological contact.

Nowadays, there is only one standard related to tribocorrosion tests in aqueous media, namely 'ASTM G119-09 Standard Guide for determining synergism between wear and corrosion'. But this standard has some technical drawbacks and cannot be applied to all kinds of materials. Because of that, several authors have developed their own testing procedure by performing simultaneously tribological and electrochemical measurements and by using, in many cases, homemade adapted devices. Details on tribocorrosion techniques can be found in Refs 23–30.

8.4.2   Standards of interest to medical field

There are several testing standards related to metallic materials used in the medical field. Some of them are cited below:

- ISO Standard 14242: 'Wear of total hip-joint prostheses'
  Part 1:   Loading and displacement parameters for wear-testing machines and corresponding environmental conditions for test
  Part 2:   Methods of measurement
  Part 3:   Loading and displacement parameters for orbital bearing type wear testing machines and corresponding environmental conditions for test.

- ISO Standard 14243 (also ASTM F2083): 'Wear of total knee-joint prostheses'
  Part 1:   Loading and displacement parameters for wear-testing machines with load control and corresponding environmental conditions for test
  Part 2:   Methods of measurement
  Part 3:   Loading and displacement parameters for wear-testing machines with displacement control and corresponding environmental conditions for test.

- ASTM F543-07e1 'Standard specification and test methods for metallic medical bone screws'. This specification provides requirements for materials, finish and marking, care and handling, and the acceptable dimensions and tolerances for metallic bone screws that are implanted into bone.
- ASTM F136-08e1 'Standard specification for wrought titanium-6 aluminum-4 vanadium ELI (extra low interstitial) alloy for surgical implant applications (UNS R56401)'. This specification covers the chemical, mechanical, and metallurgical requirements for wrought annealed titanium-6aluminum-4vanadium ELI (extra low interstitial) alloy (R56401) to be used in the manufacture of surgical implants. The products are classified into: strip, sheet, plate, bar, forging bar, and wire.
- ASTM F2063-05 'Standard specification for wrought nickel–titanium shape memory alloys for medical devices and surgical implants'. This specification covers the chemical, physical, mechanical, and metallurgical requirements for wrought nickel–titanium bar, flat rolled products, and tubing containing nominally 54.5% to 57.0% nickel and used for the manufacture of medical devices and surgical implants.
- ASTM F1713-08 'Standard specification for wrought titanium-13niobium-13zirconium alloy for surgical implant applications (UNS R58130)'. This specification covers the chemical, mechanical, and metallurgical requirements for wrought titanium-13niobium-13zirconium alloy to be used in the manufacture of surgical implants.

- ASTM F1377-08 'Standard specification for cobalt-28chromium-6molybdenum powder for coating of orthopaedic implants (UNS R30075)'. This specification covers the requirements for cobalt–28chromium–6molybdenum alloy powders for coating of orthopaedic implants. This specification covers powder requirements only and does not address coating properties.
- ASTM F688-05 'Standard specification for wrought cobalt-35nickel-20chromium-10molybdenum alloy plate, sheet, and foil for surgical implants (UNS R30035)'. This specification covers the requirements for wrought cobalt-35nickel-20chromium-10molybdenum alloy for use in the manufacture of surgical implants.
- ASTM F1537-08 'Standard specification for wrought cobalt-28chromium-6molybdenum alloys for surgical implants (UNS R31537, UNS R31538, and UNS R31539)'. This specification covers the chemical, mechanical, and metallurgical requirements for three wrought cobalt–28chromium–6molybdenum alloys used for surgical implants.
- ASTM F138-08 'Standard specification for wrought 18chromium-14nickel-2.5molybdenum stainless steel bar and wire for surgical implants (UNS S31673)'. This specification covers the requirements for wrought 18chromium-14nickel-2.5molybdenum stainless steel bar and wire used for the manufacture of surgical implants.
- ISO 10993 series:

  - ISO 10993-1:2003 Biological evaluation of medical devices – Part 1: Evaluation and testing
  - ISO 10993-2:2006 Biological evaluation of medical devices – Part 2: Animal welfare requirements
  - ISO 10993-3:2003 Biological evaluation of medical devices – Part 3: Tests for genotoxicity, carcinogenicity and reproductive toxicity
  - ISO 10993-4:2002/Amd 1:2006 Biological evaluation of medical devices – Part 4: Selection of tests for interactions with blood
  - ISO 10993-5:2009 Biological evaluation of medical devices – Part 5: Tests for in vitro cytotoxicity
  - ISO 10993-6:2007 Biological evaluation of medical devices – Part 6: Tests for local effects after implantation
  - ISO 10993-7:1995 Biological evaluation of medical devices – Part 7: Ethylene oxide sterilisation residuals
  - ISO 10993-8:2001 Biological evaluation of medical devices – Part 8: Selection of reference materials
  - ISO 10993-9:1999 Biological evaluation of medical devices – Part 9: Framework for identification and quantification of potential degradation products
  - ISO 10993-10:2002/Amd 1:2006 Biological evaluation of medical devices – Part 10: Tests for irritation and delayed-type hypersensitivity
  - ISO 10993-11:2006 Biological evaluation of medical devices – Part 11: Tests for systemic toxicity
  - ISO 10993-12:2007 Biological evaluation of medical devices – Part 12: Sample preparation and reference materials (available in English only)
  - ISO 10993-13:1998 Biological evaluation of medical devices – Part 13: Identification and quantification of degradation products from polymeric medical devices

- ISO 10993-14:2001 Biological evaluation of medical devices – Part 14: Identification and quantification of degradation products from ceramics
- ISO 10993-15:2000 Biological evaluation of medical devices – Part 15: Identification and quantification of degradation products from metals and alloys
- ISO 10993-16:1997 Biological evaluation of medical devices – Part 16: Toxicokinetic study design for degradation products and leachables
- ISO 10993-17:2002 Biological evaluation of medical devices – Part 17: Establishment of allowable limits for leachable substances
- ISO 10993-18:2005 Biological evaluation of medical devices – Part 18: Chemical characterisation of materials
- ISO/TS 10993-19:2006 Biological evaluation of medical devices – Part 19: Physico-chemical, morphological and topographical characterisation of materials
- ISO/TS 10993-20:2006 Biological evaluation of medical devices – Part 20: Principles and methods for immunotoxicology testing of medical devices.

8.5   Standardisation aspects and reporting on test data

The characteristics of any tribological contact vary accordingly to the imposed conditions. If a close simulation of a practical wear situation is desired, then it is necessary to characterise the tribological contact fully. Characterisation means that all major controlling factors are identified and given a suitable value. There are a large number of controlling factors, for example, the load on a contact is also related to parameters such as contact stress and time variation of load (if any). Similar considerations apply to temperature, sliding speed, etc. In order to achieve a close match between test and real tribological contact, the operational, material, environmental and lubricant parameters should be fully characterised.

Factors affecting the tribocorrosion process and mechanisms have to be taken into account for a standard definition, namely:

- The properties of the contacting materials: The microstructure of the materials and the presence of defects such as phase distribution, grain size and orientation, non-metallic inclusions, segregations, dislocation density, etc. are critical for the mechanical behaviour of the materials.
- The mechanics of the tribological contact: The rate of tribocorrosion for a given metal–environment combination depends on the applied forces and the type of contact sliding, fretting, rolling or impact. The other factors include sliding velocity, type of motion, shape and size of contacting bodies, alignment, vibration, etc.
- The physico-chemical properties of the environment: Its influence is in the form of the medium at the interface, i.e. solid, liquid or gaseous and its corresponding properties such as viscosity, conductivity, pH, corrosivity, temperature, etc.

A summary of the main parameters necessary in the characterisation of tribological contacts is outlined in Figure 8.2. This list of parameters is not exhaustive as there will always be some other specialised parameters for specific experiments. The large number of parameters involved has two consequences:

- The probability of achieving good comparability without an adequate characterisation of materials, environment and operating conditions is negligible.

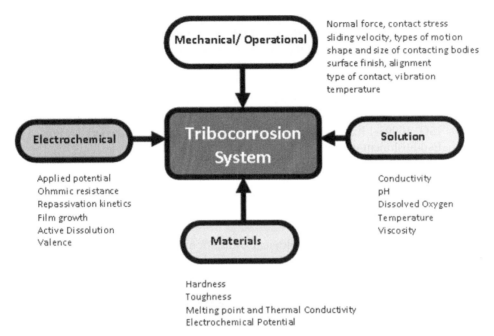

Mechanical/ Operational

Normal force, contact stress
sliding velocity, types of motion
shape and size of contacting bodies
surface finish, alignment
type of contact, vibration
temperature

Electrochemical

Tribocorrosion System

Solution

Applied potential
Ohmmic resistance
Repassivation kinetics
Film growth
Active Dissolution
Valence

Conductivity
pH
Dissolved Oxygen
Temperature
Viscosity

Materials

Hardness
Toughness
Melting point and Thermal Conductivity
Electrochemical Potential

8.2 Commonly used parameters in the characterisation of tribological contacts

- A comprehensive test programme covering all variables would require an impossibly large research effort, hence the data available from the experiments provide only fragmentary information.

It is within this structure (comprising the parameters listed in Figure 8.2) containing a very large number of degrees of freedom that research must be carefully planned to obtain meaningful data.

8.6   Evolution and convergence of actions towards standardisation of tribocorrosion carried out at the European level

8.6.1   ENIWEP – Tribostand

The Tribostand project is framed under the Eureka ENIWEP Umbrella (www.eureka.org), with the ultimate aim to ensure efficient transfer of tribotechnology to the European economic fabric by generating European projects.

Objectives

The main objective of the Tribostand project consisted of carrying out a complete revision of the current laboratory tests which are frequently being used to characterise the tribological behaviour of metallic and ceramic materials. This project had the intent to establish some common testing protocols with all project partners and, when all conditions were fulfilled, they could be sent to suitable committees to standardise them.

The main aim of this project is to benefit companies by providing them with a complete tribological analysis service carried out by all project partners, which will be available to all companies in different sectors. In this way, companies will be able to choose whatever centre they want depending on the market they belong to and their own specialisation.

If the objectives are achieved, it will contribute to eliminate technical frontiers among European research centers and companies. It will establish testing procedures in the field of tribocorrosion which could be standardised, e.g. by CEN (European Committee for Standardization). Moreover, Tribostand will improve transference of technology within the scope of tribology (neither between participant centres in the project nor between that they are not it) and it will facilitate the transference of products in the frame of the European Union.

Technical description

Numerous test methods are currently available for studying the characteristics related to the friction and wear of materials. Quite often, the same property is measured by different equipment or methods and whose results cannot be compared. On other occasions, it is necessary to design a specific apparatus to study a particular problem. However, the test results obtained with this apparatus cannot then be extrapolated to other materials. In order to address this problem, the most frequently performed tribology tests will be compiled and, wherever possible, a single method will be adopted for each test, based on the results of round-robin inter-comparison tests, study of the variables, etc., to extrapolate the results obtained in the different technology centres or company laboratories.

Various standardisation actions are currently being undertaken in subjects related to the friction and wear of materials. A number of ENIWEP project partners are actively participating in these actions, as members of the following Standardisation Committees on areas related to tribology:

- CEN TC184 WG 5: Methods of Test for Ceramic Coatings
- CEN BT WG 166: Nanotechnologies
- ISO TC107 SC2: Test Methods for Metallic and Inorganic Coatings
- ISO TC 229: Nanotechnologies.

The participation of consortium members on these committees helps to collate the tribological tests and the material means used to conduct these tests, and to enable harmonisation of the criteria among the different Centres and companies participating in the project. This is expected to yield a group of common test protocols, which will allow the same results to be achieved, under the same conditions, at any laboratory that has the required means. Once these protocols have been obtained and their validity and traceability verified, a tribological analysis and test service will be established which will enable companies from different sectors to solve their problems at any member Centre of the service; the choice of Centre will only depend on the particular needs of each company.

The consortium is composed of:

- Research groups specialised in tribology and interested in harmonising their tests and/or taking part in the mentioned tests service.
- Equipment manufacturers interested in developing or adapting their equipment according to the standardised tests.
- Companies whose products are subjected to wear and friction processes, interested in taking part in test procedures and establishment of a test service.

The final results of the project are supposed to be applied directly to all of the industrial sectors whose products are subject to wear and friction processes.

The companies from these sectors will benefit from the analysis and test services resulting from this project.

The Tribostand project has the aim of creating the following new services, products and standard test methods:

- Services
  - New services of advanced characterisation of coatings
  - New services in tribological evaluation of materials and coatings
  - New customised services of tribological tests depending on intended use.

- Products
  - New tribological testing machines able to work with new standards
  - New characterisation machines
  - Harmonised test methods
  - Standard test methods.

According to these products obtained in the project, the research centres will be responsible for the exploitation of services and standardisation methods, while test machine manufacturers will be in charge of the development and adaptation of testing machines.

With a view to providing some indications of the foreseen impact, note that approximately 25% of global energy is lost through friction. Losses owing to wear of mechanical parts are estimated at 1.3–1.6% of GDP in industrialised countries. The costs associated with friction, wear and lubrication problems are estimated at approximately 350 billion EUR per annum.

The principal tribo-sensitive sectors are as follows: land transport (46.6%), industrial processes (33%), energy supply (6.5%) and aeronautics (2.8%).

8.6.2 COST 533 Biotribology

COST Action 533 Biotribology aims to study 'Materials for improved wear resistance of total artificial joints'. This Action is framed in the Materials, Physical and Nanosciences Domain Committee, whose purpose is to ensure efficient transfer of biotribotechnologies between different research centres and consequently to lead to future collaborations.

Objectives

The main objective of the Action was to develop materials for improved wear resistance of artificial joints and novel low wearing designs. For this purpose, the biotribology of materials in artificial hip and knee joints, the mechanisms of lubrication and wear, the methods of *in vitro* simulation and testing, and the resulting biocompatibility and biological reactions to the wear products were investigated to contribute to the standardising of *in vitro* simulation and testing. This development furthermore resulted in reduced numbers of wear particles, lower adverse biological reactions and longer lifetimes of artificial joints and natural joints.

Technical description

The main scope of this action was to investigate the biotribology of bearing materials in artificial hip and knee joints, the mechanisms of lubrication and wear, the methods of *in vitro* simulation and testing, and the resulting biocompatibility and biological reactions to the wear products to contribute to the standardising of *in vitro* simulation and testing. Three major initiatives were undertaken, including a pan-European study of highly cross-linked polyethylene for hip joints, a round-robin tribocorrosion testing of metallic bearing materials for metal-on-metal hip joints, and fundamental studies of the lubrication mechanism in natural synovial joints and interventions, including a tissue engineering approach.

The working groups are:

WG 1: Enhanced and cross linked polyethylene in artificial hip and artificial hip and knee joints.
Research themes: Mechanical properties and fatigue; Kinematics, conditions and wear; Abrasive and third body wear; Lubricants and lubrication; wear debris and biocompatibility.
WG 2: Alternative hard bearing couples for artificial hip joints.
Research themes: Fracture and fracture toughness; Joint laxity and surgical constraint; Lubrication content and conditions; Design and geometry; Wear debris and biocompatibility.

A testing protocol on tribocorrosion and its validation was carried out, and an experimental round-robin test action entitled 'Tribocorrosion methods for the characterisation of CoCrMo biomedical alloys in simulated body fluids'.

WG 3: Wear of polyethylene in artificial knee joints
Research themes: Contact mechanics and stress analysis and fatigue; Kinematics and joint simulation; Surface modifications and engineering; Surgical conditions and procedures; Backside wear.
WG 4: Therapies and treatments to extend the life of the natural joint.
Research themes: Degradation and degeneration characterisation and modelling; Lubricants and lubrication mechanisms; Therapeutic lubricants and gels; Matrix substitutions (Natural cartilage and cartilage substitution using novel coatings, Novel cartilage substitution hydrogel materials, Bone regeneration and improved cartilage formation in the body at ectopic locations); Regenerative medicine; Tissue engineering.

The outcome from such studies provides timely guidelines for the pre-clinical evaluation of these treatment options, and provides knowledge for the regulatory and standards bodies. This will have economic benefits to health services as well as clinical benefits to patients. Furthermore, significant networking within Europe as well as with other countries, such as China, USA, Japan, Australia and Israel, has been established. A number of early stage researchers (particularly female researchers) have been trained and developed throughout the Action. All of these will contribute to the establishment of a centre of excellence in biotribology in Europe.

During the COST 533 Action, a variety of workshops, conferences and training schools have taken place with subsequent publications and reports, resulting in the exchange of findings obtained in research studies, as well as the possibility of making Short-term Scientific Missions in other centres or laboratories.

8.7   Conclusions

The need for standardisation of both research and industrial practices has been highlighted both in terms of technical and economic benefits. The standardisation process requires different steps from the technical description of protocols to the final publication of the international standards. There is an interest in standardisation protocols in the world of ceramics and metals, but since metals are more prone to tribocorrosion processes, the most important effort has been focused on metals and protective coatings. The ASTM G119-09 Standard is a guide for determining the synergism between wear and corrosion. The tribocorrosion process is very typical in biomaterials where composition, wear behaviour and biocompatibility are covered for a large number of ASTM and ISO standards.

The Eureka Umbrella ENIWEP (www.eureka.org) launched the European Project Tribostand to establish some testing protocols in tribology and tribocorrosion. The work carried out in this action has been coordinated by other European Standardisation Committees CEN TC184 WG 5 ('Methods of tests for ceramic coatings'), CEN BT WG 166 and ISO TC 229 ('Nanotechnologies'), ISO TC107 SC2 ('Test methods for metallic and inorganic coatings'). The COST 533 Action on Biotribology 'Materials for improved wear resistance of total artificial joints' has the objective to ensure efficient transfer of biotribotechnologies between different research centres. In the frame of this action, there is a group working on alternative hard bearing couples for artificial hip joints. One of the activities pushed by this working group was the elaboration of a testing protocol on tribocorrosion and its validation through a round-robin at the European level, and which sought the participation of 16 laboratories in the field of tribocorrosion applied to the field of prosthetics.

References

1. I. Greenway: Why Standardise? In: 'Promoting Land Administration and Good Governance', 5th FIG Regional Conference, Accra, Ghana, 8–11 March 2006.
2. W. K. Buck: Report on National Materials Policy Conference at Henniker, Mineral Resources Branch, 1970.
3. G. W. Stachowiak and A. W. Batchelor: 'Engineering tribology'; 2006, Amsterdam, Elsevier. ISBN 978-0-7506-7836-0.
4. P. Jost: Lubrication (Tribology) - Education and Research: A Report on the Present Position and Industry Needs, Publ. Department of Education and Science, HM Stationary Office, London, 1966.
5. Research Report (T76-38) Tribology (Code BMFT-FBT76-38), Federal Ministry for Research and Technology, West Germany, 1976.
6. 'ISO Stages of the development of International Standards'. http://www.iso.org/iso/standards_development/processes_and_procedures/stages_description.htm. Geneva, ISO. Date last accessed 31 December 2009.
7. 'ISO/IEC Directives, Part 1 – Procedures for the technical work', Sixth edition; 2008, Geneva, ISO. Date last accessed 1 January 2010.
8. 'ISO/IEC JTC 1/SC 29, SC 29/WG 11 Structure (ISO/IEC JTC 1/SC 29/WG 11 – Coding of Moving Pictures and Audio)'; 2009. Geneva, ISO. Date last accessed 7 November 2009.
9. 'International harmonized stage codes'. http://www.iso.org/iso/standards_development/processes_and_procedures/stages_description/stages_table.htm. Geneva, ISO. Date last accessed 31 December 2009.
10. 'The ISO27k FAQ – ISO/IEC acronyms and committees'. http://www.iso27001security.com/html/faq.html#Acronyms. IsecT Ltd. Date last accessed 31 December 2009.

11. 'US Tag Committee Handbook'. 2008-03. Date last accessed 1 January 2010.
12. Letter Ballot on the JTC 1 'Standing Document on Technical Specifications and Technical Reports'. Geneva, ISO. Date last accessed 1 January 2010.
13. ISO/IEC Guide 43-1:1997 Proficiency testing by interlaboratory comparisons – Part 1: Development and operation of proficiency testing schemes. Geneva, ISO.
14. ISO/IEC Guide 43-2:1997 Proficiency testing by interlaboratory comparisons – Part 2: Selection and use of proficiency testing schemes by laboratory accreditation bodies. Geneva, ISO.
15. S. Jahanmir: *Proc. Inst. Mech. Eng. Part J: J. Eng. Tribol.*, 2003, **216**, 371–385.
16. J. R. Gomes, A. S. Miranda, R. F. Silva and J. M. Vieira: *J. Am. Ceram. Soc.*, 1999, **82**, (4), 953–960.
17. T. E. Fisher and W. M. Mullins: *J. Phys. Chem.*, 1992, **96**, 5690–5710.
18. A. Tucci and L. Esposito: *Wear*, 1994, **172**, 111–119.
19. A. Conde, C. Navas and J. J. Damborenea: *Rev. Metal. Madrid Vol. Extr.*, 2005, 457–462.
20. C. Monticelli, F. Zucchi and A. Tampieri: *Wear*, 2009, **266**, 327–336.
21. C. Richard, C. Kowandy, J. Landouisi, M. Geetha and H. Ramasawmy: *Int. J. Refract. Met. Hard Mater.*, 2010, **28**, 115–123.
22. R. J. K. Wood, D. Sun, M. R. Thakare, A. Frutos Rozas and J. A. Wharton: *Tribol. Int.*, 2010, **43**, 1218–1227.
23. S. Mischler and P. Ponthiaux: *Wear,* 2001, **248**, 211–225.
24. I. García, D. Drees and J. P. Celis: *Wear,* 2001, **249**, 452–460.
25. D. Landolt, S Mischler and M. Stemp: *Electrochim. Acta,* 2001, **46**, 3913–3929.
26. P. Jemmely, S. Mischler and D. Landolt: *Wear,* 2000, **237**, 63–76.
27. P. Ponthiaux, F. Wenger, D. Drees and J. P. Celis: *Wear,* 2004, **256**, 459–468.
28. J. P. Celis, P. Ponthiaux and F. Wenger: *Wear,* 2006, **261**, 939–946.
29. M. Azzi and J. A. Szpunar: *Biomol. Eng.*, 2007, **24**, 443–446.
30. A. J. Sedriks: 'Corrosion of stainless steel'; 1996, New York, Wiley.

abrasive wear 72–3
acoustic emission (AE) 40, 49–50
'active wear track' concept 5, 123–4
adhesive wear 72
adsorption 7–8, 109
amorphous carbon coatings 77
antagonist materials 126
articulating systems 14
ASTM standards 191–2, 198
atomic emission spectroelectrochemistry
    (AESEC) 32
atomic force miscroscopy (AFM) 74

ball-on-flat configuration in tests of wear 59–60
bare metal surfaces, electrochemical phenomena
    on 33–7
bearing materials 67
bench tests 120
biomaterials and biocompatibility 17–19

cathodic polarisation 151
cavitation wear 74
channel flow double electrodes (CFDEs) 42–4
characterisation of tribological contacts 193–4
chemical industries 25
chemical mechanical polishing (CMP) 24
classical approaches to tribocorrosion 110–11
co-based alloys 18
coefficient of friction 69–70, 96–7, 110
combinatorial testing 84
contact pressure in a tribological system 80, 83
core standards 10
corrosion, definition of 3
corrosion and wear, interaction between 95–7;
    see also synergy
corrosion wear 74
COST 533 Biotribology 196–8
crevice corrosion 37
cutting tools 25

degradation of materials 71, 88
dental implants 22–4
depassivation 29, 36

economics
    of standardisation 185
    of tribology and tribocorrosion 187

elastohydrodynamic (EHL) conditions 58
electrochemical impedance spectroscopy (EIS)
    101–9, 168–70, 173
electrochemical methods for tribocorrosion
    studies 125
electrochemical microcell techniques 114–15
electrochemical noise (EN) analysis 109–10
electrochemical properties of materials 29–38
electrochemical responses 50, 122, 168–70
entry angle in a tribological system 80–1
erosion wear 73–4
European Union, standardisation within 185,
    194–7
extreme pressure (EP) performance of lubricants
    63, 65

Faradaic balance 31, 52
Faraday's law 40, 48–9, 142
fatigue corrosion 37
fatigue wear 74
fretting tests 8, 77
friction and wear
    factors influencing 56–7
    interpretation of test data on 61–3
    relationship between 77–8, 85
    scale dependence of 69–71
    tribological conditions for 115
frictional forces 55, 61, 72, 120–1
fuel efficiency 188
galvanic coupling 90–3, 98–102, 111, 116, 121,
    151, 164–6
galvanic effects 44–7
Garcia, I. 89
Greenwood, J.A. 71

Hanniker Conference on National Material Policy
    186
'high field conduction' type models 130
hip-joint prostheses 20–2
Holm-Archard equation 113

implants *see* dental implants; orthopaedic
    implants
inductively coupled plasma optical emission
    spectrometers (ICP-OESs) 32
insulation in tribocorrosion test systems 189
international standards 188–93

Jemmely, P. 89
Jost, Sir P. 186

laser induced decohesion technique (LIDT) 43
latency times 151, 158–62
lithography, use of 112
losses due to friction and wear 2, 6, 188
lubrication 25
    factors influencing 57–8
    interpretation of test data 63
    materials used for 57

material degradation
    causes of 2
    data analysis on 172–4
    during tribocorrosion tests 6
    effects of 26
mechanical breakdown, simulation of 40–1
mechanical film-rupture 38–40
medical standards 191–2
meso-tribometers 68–9
microelectrodes 116
microprobes 114

normalisation of tribocorrosion 10–11
nuclear sector 25–6

open circuit potential (OCP) measurements 90–4,
    100–1, 106–9, 116, 169–72
organisation standards 10
orthopaedic implants 20
oxidation 73, 131–4, 140–2

passivation 30–33, 89–90
passivity 29, 88
    breakdown of 37–8
pitting corrosion 37, 100
plasticity index 71
polarisation curves 94–8, 101
'potential steps' experiments 98–100
potentiodynamic polarisation measurememts
    94–7
profilometry 74–5, 85, 147

reciprocating unidirectional sliding tests 8
Rehbinder effect 15
repassivation 29, 36, 38, 41–4, 52, 158–62
repassivation current transients 99–101
rotating ring-disc electrodes (RRDEs) 31–2, 42
'running in' of materials 84

scanning electrochemical microscopy
    (SECM) 113–14
scanning vibrating electrode technique
    (SVET) 44–6
shear strain 111
'sliding tests' 135–40, 152–63, 167

continuous and intermittent conditions for
    8, 129, 143, 147, 152–3
    results of 160–3
    rotating and reciprocating 170
slurry erosion–corrosion experiments 40–1
specific polarisation resistance 102
specification standards 10
Stack, M.M. 150
stainless steel 18, 89–98, 105–8
    case study on the calculation of tribo-corrosion
        components 174–83
standardisation
    definition of 184
    economics of 185
    need for 184–5, 198
    process of 188–9
    of test methods 61–3
    upstream and downstream 11
standards
    for ceramic and metallic materials 186–91
    definition of 9, 184
    medical 191–2
    related to tribology 10
strain 111–12
straining electrode theory 39–40
stress corrosion cracking 37, 100
Stribeck curves 57, 59
synergy between corrosion and wear 5, 47–52,
    121–4, 129, 150–66
    comparison between previous and new
        approaches 163–4
    design of a new test protocol 152–60, 166
    previous approaches to 150–2

thin later activation (TLA) 85
thin microelectronic coatings 67
titanium alloys 18
tribocatalysis 7
tribochemistry 7
tribocorrosion 2–7, 79, 88–9, 100–1, 187–8
    causes and results of 6
    concept and definition of 2–5, 14–15
    in industrial applications 24–6
    in medical applications 16–24
    multi-scale approaches to 110–15
    parameters affecting 15–16
    protocol on 11
    resistance to 122
    scope of 2
    simplified model of 129–46
    standard definitions of processes and
        mechanisms 193
tribocorrosion experiments, design of 119
tribocorrosion systems 4, 120–2
tribocorrosion testing 4, 6, 8–11, 101–6, 109, 112,
    115–16
    general principles of 127–9
    lack of standard apparatus for 189–90

measurements and techniques used in 146–8
and normalisation 10–11
and oxidation 140–2
selection of parameters for 167–8
specificity of 8–9
tribological aspect numbers (TANs) 78–9, 81–2
tribological testing, choice of conditions for
    126–7
tribology 1, 3
tribology equation 186
tribology test equipment 63–71
triboluminescence 7
tribometers 63–8, 84–5, 89, 110–13, 116
    choice of 124–6
    classical 110–11
    classification of 65–8
    combined with microelectrodes 116
    displacement modes of 68
    load ranges of 67
    modular 64–5
    pin-on-disk type 116, 125–8
    reciprocating type 125–6
    sensitivity of 68
tribometrics 83
tribo-oxidation 7
triboreactivity 7
Tribostand project 11, 194–8
tribosystems 78

tribotesting
    accelerated 83–4
    simultaneous 84
    strategy for 78
    see also tribocorrosion testing

ultramicroelectrodes 114

Watson, S.W. 88–9, 150
wear
    in automotive engines 59
    characterisation methods 71–8
    definition of 55
    in relation to operating conditions 187
    see also friction and wear
wear components, calculation of 142–5
wear loss 61
wear modes 72–4
wear rates
    absolute and *relative* 84
    quantification of 74–7
Williamson, J.B.P. 71
Working Group on TriboCorrosion (WP18) 11

X-ray photoelectron spectroscopy (XPS) 29

zero resistance ammeters (ZRAs) 47

Printed and bound by CPI Group (UK) Ltd, Croydon, CR0 4YY

21/10/2024

01777040-0007